CRITIQUES

Studies in Critical Social Sciences Book Series

Haymarket Books is proud to be working with Brill Academic Publishers (www.brill.nl) to republish the *Studies in Critical Social Sciences* book series in paperback editions. This peer-reviewed book series offers insights into our current reality by exploring the content and consequences of power relationships under capitalism, and by considering the spaces of opposition and resistance to these changes that have been defining our new age. Our full catalog of *SCSS* volumes can be viewed at https://www.haymarketbooks.org/series_collections/4-studies-in-critical-social-sciences.

Series Editor
David Fasenfest (York University, Canada)

Editorial Board
Eduardo Bonilla-Silva (Duke University)
Chris Chase-Dunn (University of California–Riverside)
William Carroll (University of Victoria)
Raewyn Connell (University of Sydney)
Kimberlé W. Crenshaw (University of California–LA and Columbia University)
Raju Das (York University, Canada)
Heidi Gottfried (Wayne State University)
Alfredo Saad-Filho (Queen's University Belfast)
Chizuko Ueno (University of Tokyo)
Sylvia Walby (Royal Holloway, University of London)

CRITIQUES

In Defence of Development

TOM BRASS

Haymarket Books
Chicago, IL

First published in 2024 by Brill Academic Publishers, The Netherlands
© 2024 Koninklijke Brill NV, Leiden, The Netherlands

Published in paperback in 2025 by
Haymarket Books
P.O. Box 180165
Chicago, IL 60618
773-583-7884
www.haymarketbooks.org

ISBN: 979-8-88890-556-2

Distributed to the trade in the US through Consortium Book Sales and Distribution (www.cbsd.com) and internationally through Ingram Publisher Services International (www.ingramcontent.com).

This book was published with the generous support of Lannan Foundation, Wallace Action Fund, and the Marguerite Casey Foundation.

Special discounts are available for bulk purchases by organizations and institutions. Please call 773-583-7884 or email info@haymarketbooks.org for more information.

Cover design by Jamie Kerry and Ragina Johnson.

Printed in the United States.

Library of Congress Cataloging-in-Publication data is available.

*For Amanda,
Anna, Ned, and Miles;
and in memory of my parents*

Contents

Acknowledgements XI

Introduction: Last Rites for Development Studies? 1
Are We the Masters Now? 5
Dismantling Development 7
Post-development? 9
This Sense of Identity 11
Climate, Class, Risk 14
Themes 17

PART 1
Questioning the Paradigm

1 **Racism and Development: Blood, Sweat and Fears** 25
Introduction: More Lessons from History 25
An Absent Nationality 27
A Race against Time, a Time against Race 30
Southern Myths 34
No Ear to Hear 36
Differences, Sameness 39
Common Heirs to Its Impositions 42
The Battle for Bread 44
The Pinch of Hunger 46
Conclusion 49

2 **The Industrial Reserve and Development: a Vanishing Army?** 52
Introduction: Redefining the Industrial Reserve 52
19th Century Marxist Views 55
20th Century Liberal Views 60
20th Century Marxist Views 63
Border Wars 68
Human Flourishing, but Whose? 72
What Marx Really Said 76
Travelling the Same Road? 78
Conclusion 81

3   **Sociology and Development: a Warning from *The History Man***   84
    Introduction: Publishing, Hierarchy, Power   84
    The Bleak End of Things   85
    Who Is *The History Man* Now?   90
    The Power of Hierarchy   93
    No One Is Listening?   96
    Conclusion   98

4   **Critical Agrarian Studies and Development: a Populist Land Grab**   101
    Introduction: the Sleep of Forgetfulness   101
    In the Academic Salon   105
    Deprivileging Marxism   106
    'Marxist' Questions   109
    'Marxist' Answers   111
    Reprivileging Agrarian Populism   114
    Conclusion   117

PART 2
Alternative Agendas

5   **Development: a Theory without a Past, Present, or Future?**   121
    Introduction: Paradigms/Concepts That Disappear/Reappear   121
    Call a Friend   123
    Concepts, Origins   126
    Capitalism Everywhere, Capitalism Nowhere   128
    Development Theory?   131
    The Sharpest Weapon   137
    Conclusion   141

6   **Liberalism and Development: Fukuyama's Scylla and Charybdis**   143
    Introduction: a Benign Capitalism?   143
    Floreat Classical Liberalism?   148
    I Am a Nice Shark …   150
    A Progressive Left?   153
    Political Corrections, Problematic History   155
    Conclusion   159

7   **Anthropology and Development: Self in the World, World in the Self**   161
     Introduction: What Do I Know?   161
     The Self (in the World)   164
     Self-Help   166
     No Friends There   168
     The World (in the Self)   171
     Insufficiency   173
     Self-Sufficiency   174
     Humanity's Priority   176
     Restlessness   178
     Conclusion   180

8   **Labour Regime and Development: Deproletarianisation and Neo-bondage Compared**   182
     Introduction: Explaining Unfree Labour   182
     Deproletarianisation, Neo-bondage   184
     Unfreedom, Patronage, Politics   187
     Differences Explained?   191
     Misinterpreting Capitalism   194
     Conclusion   196

PART 3
Beyond Capitalism?

9   **Postmodernism and Development: Misremembering the Peasantry**   201
     Introduction: Doing without Development?   201
     Methodology   204
     Sources   207
     Stories   208
     Theory   211
     Definitions   211
     Economy   213
     Politics   215
     Conclusion   217

10   **On the Continuing Necessity of (Marxist) Critique**   219
     Introduction: Paradigms, Polemics, Popularity   219
     A Return to Yesterday   222
     New Paradigms, Old Assumptions   223
     Class Dismissed   225
     Producing Curtains   226
     Urgent Need of Renewal   229
     Conversation, Collaboration, Cooperation?   231
     Hegemonic Formation, Populist Moments, Floating Signifiers?   234
     Taking People's Beliefs Seriously?   236
     Conclusion   237

Conclusion   239

Bibliography   249
Author Index   269
Subject Index   273

Acknowledgements

Forgetting that systemically there are different paths of economic development, much discussion in the social sciences has nevertheless concluded that as all forms of development are politically suspect – categorizing them as either having failed, or not being environmentally feasible – such a process is no longer viable anywhere. In a large part, this kind of negative assessment can be traced to two causes: first, the role of industrialization in generating climate change; and second, the antagonism expressed by postmodern theory towards what it terms foundational Eurocentric approaches that privilege unwarranted economic growth. Among the condemned development paradigms is Marxism, dismissed for its emphasis on production, a result of planning being associated by many as much with capitalism as with socialism. Similarly untoward is an additional conflation, licensing the widespread misrepresentation of culture wars as being waged by Marxism, instead of against it.

Saying that Marxist theory based on class is no longer a valid method of thinking about development quickly moves onto the assertion by postmodernism that it never was. An entirely predictable effect of this epistemological shift is the creation of a space in which the main alternative mobilizing discourse – non-class identity politics – comes to the fore and thrives in academic circles. Not the least problematic result of the wholesale annexation of development issues by postmodernism is the privileging of national/ethnic discourse, both for and against, which feeds into how the industrial reserve army of labour is perceived, a central determinant of the rise and consolidation in metropolitan capitalist nations of populism. These processes have been consecrated as academic fashion and reinforced by its form of institutional hierarchy, a locus where an entirely predictable outcome of delegitimizing Marxist development theory based on class has facilitated the broader critique by political economy of class being replaced by the postmodern celebration of culture.

Special thanks are due to the following people. To Professor David Fasenfest, the Series Editor, for encouragement; to Judy Pereira of Brill Publishers, who guided the book through production; and to my daughter Anna Luisa Brass, who provided the drawings for the front cover and those within the book itself.* She drew the cover for seven of my previous books – *New Farmers' Movements in India* (1995), *Labour Regime Change in the Twenty-First Century*

---

* More artwork by Anna can be found in her forthcoming book *Eyes are holes in your head* (2024).

(2011), *Class, Culture and the Agrarian Myth* (2014), *Labour Markets, Identities, and Controversies* (2017), *Revolution and Its Alternatives* (2018), *Marxism Missing, Missing Marxism* (2021), and *Transitions* (2022).

A number of chapters draw on materials which have appeared previously in different journals. Others have not been published before, and appear here in print for the first time. Like all my previous monographs, this one is dedicated to two sets of kin. To my family: Amanda, and Anna, Ned and Miles. Also, to the memory of my parents: my father, Denis Brass (1913–2006), and my mother, Gloria Brass (1916–2012).

Richmond-upon-Thames
May, 2024

Introduction: Last Rites for Development Studies?

Happy is he who forgets what can no longer be changed.
> A view about peace of mind secured by abrogating political engagement attributed to the nineteenth century Austrian composer JOHANN STRAUSS THE YOUNGER, not so different from the one held now by social scientists engaged in the study of development.

∴

Why this and not that?
> A simple question about film language and scripts posed by DAVID MAMET, a basic one that is either not asked by development theorists, or when asked endlessly perplexes them.[1]

∴

Simply put, development combines three processes: economic growth, or the generation of surplus; ideology, or who is entitled to its fruits, and why; and politics, or the realization of this entitlement. Based on different kinds of social identity (among them class, ethnicity, and nationalism), the theory, methods, practice, and struggle that stem from rival interpretations concerning this entitlement are embodied in political economy. Competing entitlements are funnelled via control over the state, which oversees the functioning of the institutional framework (legislative/regulatory ordinances, property relations), resource extraction (planning, revenues, taxation), and transfers (public spending of surpluses, welfare provision). While opposition to aspects and effects of development has always been a feature of history, what seems to be new these days is the questioning of development *tout court*. The latter position can be linked to concerns that, as any and all forms of development are thought inevitably to generate undesirable consequences, not least climate change and populism, it is necessary to scale down, roll back, or indeed

---

1 For this question, see Mamet (2023: 57). Those who have tried to apply the same question to the development process often resemble nothing so much as the hapless protagonists depicted brilliantly by Jacques Tati in his films *Jour de fête* (1949) and *Les Vacances de Monsieur Hulot* (1953).

abandon economic progress altogether. Addressed in the chapters which follow, therefore, are different aspects of the development process (including market competition, racism, populism, and unfree labour) and how these have been adjusted or redefined as a result.

To a large degree, the debate in the social sciences about political economy has involved a development process based either on the capitalist market or on the socialist plan.[2] The contours of each model are familiar: mainstream economists ranging from advocates of Keynesian demand management to supporters of a *laissez-faire* approach regard the market as an efficient form of resource allocation, capable of income redistribution via the 'trickle down' effect. Equally distinct is the property relation: as against individual private ownership favoured by the capitalist market, Marxism argues for development based on collective proprietorship vested in the state. Because for many non-Marxist development theorists capitalism as a form of economic organization has the aura of historical permanence, and consequently in effect is a system from which it is impossible to exit, no sort of transcendence is conceivable or necessary. Capitalism changes, but its logic and dynamic always remain that of the market. This, too, differs from Marxist theory which, since it historicizes development, ties the latter to contradictions generated by a particular mode of production; economic and social progress is driven by class formation and struggle, a process arising from systemic crisis licensing a transition to another and socially more efficient mode.

Inescapable is the extent to which development nowadays is seen largely in non-Marxist, reformist terms, as embodied in a plethora of micro-level adjustments perceived by their advocates to be feasible – not to say irreversible – solutions to the problems addressed. The core assumption of these approaches is that negative departures from what is depicted as a basically functionalist economic system are nothing more than 'anomalies', unfortunate but unconnected to its otherwise smooth operation. Solutions are as a consequence to be sought only within capitalism, not its transcendence leading to a replacement. That such negative aspects might be, after all, central to the logic of the way the economic system is reproduced tends not to be something that holders of this reformist view accept. Hence the kind of issues that require amelioration, ranging from social exclusion to fair trade, which in turn give rise to proposals like improving welfare/human rights, stimulating participatory development,

---

[2] For overviews of the history of economic theory from these different political approaches to development, see Rubin (1979) and Seligman (1962).

inaugurating poverty reduction strategies, and promoting responsibility to distant others.³

Such goals are to be realized as small-scale projects operating within particular localities, along the lines of pro-poor initiatives, the reduction of vulnerability, fostering gender empowerment, encouraging peace-building partnerships, strengthening human security, and supporting good governance, all undertaken by NGOs and kindred organizations pursuing single issue objectives by means of indigenous knowledge.⁴ Meanwhile the capitalist system, the cause of these same development problems, remains free and largely unchallenged not just to continue the very processes required by its market driven logic (oppression, exploitation, surplus appropriation), but also to co-opt any opposition, frustrate any potential/actual challenge to its dominance, and – if necessary – to reverse any gains made by labour at its expense.

At times it seems as if the only advocates of development these days are supporters of the market. Equally, there are times when it also seems that many on the left have quietly abandoned the idea of development, being content either merely to manage capitalism more efficiently than conservatives or harking back to what is now considered (wrongly) a more benign non-capitalist past. This can be seen as a tendency by the left to be distracted from the main task, signing up to other kinds of struggle, which – however worthy – deviated from what has always been (and should always be) the central objective.⁵ The issue is not the pursuit of these kinds of struggle, but rather doing so independently, outside of (and sometimes against) a wider socialist agenda. Instances of this are claims that capitalism still has a progressive role to play in terms of

---

3   These problems, together with the following solutions, can all be found in the pages of a volume about development edited by Desai and Potter (2008).
4   See, for example, the kind of development projects covered by Sumner and Tiwari (2009). Some indigenous tribal populations reject both development and leftism tout court. This is because they 'do not regard themselves part of the class system and are often suspicious of alliances with left groups' argues Bodley (1982: 172, 205), a 'rejection of the leftist political orientation' expressed in the following manner: 'Revolutionary Marxism is committed to even further perpetuation and perfection of the very industrial process which is destroying us all'.
5   This kind of distraction can be illustrated with reference to an episode I witnessed. Whilst a student at Sussex University in the early 1970s, I attended a large meeting addressed by a freedom fighter struggling for the independence of what was then Rhodesia. Following an inspiring account by him of what was happening in the field, he was asked by one student (not me) about the possibility of going out to the field and joining the guerrillas there. With a look akin to benign irritation, he answered loudly 'no'. With the apartheid regime in Rhodesia, he said, we ourselves will deal. Addressing the whole audience, he concluded: 'your task, the way you can help us over there, in the field, is to take on those politicians and businessmen here, in your own country, those who support and sustain the regime – the enemy is at home'.

development, pro-capitalist variants of which include debate both about climate change and imperialism.

As in the case of opposition to development (see below), it is also necessary to differentiate those who support the idea of development. On the one hand, therefore, are what might be termed cheerleaders for further accumulation: these extend from neoliberals to academic commentators like Fukuyama.[6] Included in this category are those who advocate going back to a nicer form of accumulation, such as Breman.[7] Also part of this same approach are Marxists who endorse development, and insist that as capitalism has not yet run its course, it must therefore be supported in this endeavour.[8] On the other hand, there are Marxists who support development, but argue that capitalism has indeed run its course and should be replaced with socialism.[9] Even if those of us who remain on the political left have learned nothing else, therefore, one unavoidably clear issue nowadays is that there is no reality that capital does not – and will not – attempt to turn to its own advantage, and thus adjust to in order to generate profit. It is perhaps no more than a sign of the times that it is still necessary to point out that it is not the task of socialists (and progressives more generally) to assist in this endeavour, by making it easier for capitalism as a system to survive and prosper. Yet it is precisely this that academic opponents of *laissez-faire* seek to do when calling not for a transcendence of capitalism but rather only for a return to a 'nicer', more 'caring' sort of accumulation.

Coming to terms with capitalism in this fashion is evident throughout development studies. Among the more implausible arguments made by its exponents are: that the solution to *laissez-faire* is a return to a more benign pattern of accumulation; that capital can and will eradicate unfree labour; that immigration is simply an issue of human rights and citizenship, unconnected with either the industrial reserve, labour market competition, or accumulation; and that where populism is concerned, as well as a 'nasty' version there is also a 'nicer' variant, which ought to be supported politically. The folly of wishing for any pattern of social progress – let alone a universal one – without at the same time eliminating capitalism is not difficult to discern. Merely addressing the economic role that the market will play, and attempting to build

---

6 See Chapter 6, this volume.
7 See Chapter 8, this volume.
8 In this category are found those such as Byres and Utsa Patnaik, whose interpretation of development is based on the semi-feudal thesis, an approach that has been subject to critical analysis (Brass, 2018a: 105ff.).
9 Different from other forms of Marxism, this approach to development takes its lead from the theory of permanent revolution as conceptualized by Trotsky.

a development programme around this, leads inevitably to a search for benign forms of accumulation that will either support or at least fit in with progressive objectives. Leaving capitalism intact to frustrate or undermine what has been achieved can be illustrated with reference to what happened in the UK after Labour victory in the 1945 election.

Are We the Masters Now?

In defiance of history, much development theory has been premised on the assumption that, once in place, legislative ordinances licensing policies designed to combat inequality, such as income redistribution, extending trade union rights, pursuing agrarian reform, higher wages and improved working conditions, could not – and would not – be reversed. This supposition was false, in that it ignored one of the main lessons that history teaches: namely, the willingness and capacity of those with power and wealth to recover property and other resources expropriated by a politically reformist state. Overlooked thereby is that class struggle is a process waged as much from above (by capitalists, landlords, bankers, merchants) as it is from below (by workers, poor peasants). Unless this dynamic is recognized – that progress necessarily requires the economic and political eradication of ruling class power and its material base – even the limited achievement of reformist measures cannot in the end be guaranteed.

Nowhere is this more evident than in the aftermath of the 1945 election landslide obtained by the Labour Party in the UK, hailed as 'the greatest British political earth tremor since 1906', a victory that '[f]or British Socialism ... was, of course, its one great historic moment'.[10] Optimism was widespread, taking the form that 'the revolution had surely begun; and, at least in Club-land, they [the ruling class] looked anxiously at the lamp-posts'.[11] Although the 1945 election

---

10 See Howard (1964: 15, 19). A major achievement of the 1945 election was undeniably the National Health Service which, together with the foundations of the welfare state, were programmatic initiatives set out in the 1942 Beveridge Report. As outlined in Chapter 2 in this volume, however, although Beveridge perceived the object of his social programme as a cure to the blight of unemployment, he nevertheless continued to regard private enterprise and the market as positive aspects of British society, in effect providing them with a political space from which was to emerge the *laissez-faire* project some four decades later.

11 Howard (1964: 21, 22), who adds that when the ruling classes saw the political shape of the Atlee cabinet, 'those in high and traditional places felt they had been given a reprieve'. A Labour MP observed (Howard, 1964: 25) that: 'What is the use of having an orderly revolution if it turns out not to be a revolution at all?'. More than the privations occasioned

was a notable step forward for the working class, much of the apparatus of bourgeois rule and power nevertheless remained intact. Among the institutions representing the latter that continued to function much as before were the monarchy, the aristocracy, the house of lords, landownership, and the public schools. Whilst the continued importance of class difference was acknowledged, struggle based on this was seen either as a from-below phenomenon or confined largely to as-yet underdeveloped countries in the Third World. That in a core area of capitalism like Britain owners of the means of production could – and indeed at some point would – mobilize not just to defend their property rights but also to undo those gains achieved as a result of government programmes of reform was a prospect largely absent from discussion.[12]

Accordingly, in the UK during the immediate post-war era, when full employment, central planning, and the expansion of the welfare state were still considered politically uncontroversial objectives, widely perceived as desirable and feasible, conservatives and neo-liberals bided their time and consolidated their ideology. It entailed not adapting to the present but rather a process of radicalisation, in effect moving further rightwards, a trajectory embodied in promoting *laissez-faire*, deregulation, privatisation, and union busting, all

---

by 1939–45 war, it was those linked to the capitalist crisis of the late 1920s and 1930s which determined the scale of the Labour vote. Hence the 1945 election, observe McCallum and Readman (1947: 44), 'was a challenge not to the Government of the last five years, nor even to its predecessors that had governed the country since the last election: it was the total record of all the governments of the twenty years between the wars that was at stake. The Conservative party had been in power throughout the whole period [so] all the evils of this unhappy period were laid at [its] door. Their rule was popularly identified with that sense of insecurity, due mainly to fear of war and unemployment, which had dogged so many during these years. There was thus a tremendous presumption in favour of change'.

12 Concerns were indeed expressed about the capacity of individual bureaucrats to frustrate Labour government initiatives (Crossman, 1965), but the idea that at the behest of capital a future government would embark on class struggle waged from above, and attempt to eviscerate trade union protections, welfare provision, as well as privatize the utilities and denationalize/outsource industrial production, was not considered. There were exceptions, of course: both Schlesinger (1953) and Mattick (1971) warned against underestimating the propensity of capitalists not just to resist expropriation but also to dismantle any 'from below' gains. In a prescient critique that anticipated the future direction taken by postmodernism, Alasdair MacIntyre wrote of the line followed by the *New Left Review* that '[t]he danger is that one will fight a series of guerrilla engagements on cultural questions which will dissipate socialist energy and lead nowhere'. Much like subsequent postmodernists, moreover, E.P. Thompson (1960: 68) replied that '[t]here is no iron law of history, discovered by Marx or Trotsky, which establishes the priority for "industrial struggle" over all other forms of political or intellectual conflict'.

enforced by a strong state.¹³ This contrasts with the leftist response to that neoliberal project when the latter finally took power in the 1980s, which – with a few exceptions – has generally taken the form of either not challenging or indeed adopting much of the programme developed by its opponents.¹⁴ Whereas the right pursued an enhanced version of its own theory, remaining true to its principles, the left by contrast has on occasion been seduced by that of its political enemies.

Dismantling Development

The kind of process referred to here, it must be stressed, is not micro-level but macro-level development. Discussion about and advocacy of micro-level development, it is true, does continue apace in the relevant academic journals, concerned as these are with issues such as the provision of clean water and improved health care for impoverished rural populations. However, debate about fundamental systemic change in the form of macro-level grand theory has seemingly ground to a halt, and with it any consideration of a future that is not in some way still a form of capitalism. This is due in part to a combination of factors: the influence of postmodern aporia and the rise of populism; the banishment of concepts such as modernity, in particular Marxist theory about the desirability and possibility of systemic development.¹⁵ Accordingly,

---

13   The institutional locus of this rightist ideological consolidation was more often than not the think-tank, which issued periodic analyses supportive of a *laissez-faire* programme based on deregulation (see, e.g., Seldon, 1961; Harris, 1961; Hunold, 1961; Barran, Johnson, Rowland, and Cromer, 1969; Friedman, 1970; Hayek, 1973).

14   In keeping with this if-you-can't-beat-them-join-them approach, from Hobsbawm ('forward march of labour halted') via Stuart Hall ('learning from Thatcher') to Martin Jacques ('new times'), the object for many on the British left at that conjuncture became nothing less than to discard class analysis and socialism (labelled 'passé') in order then to embrace the market (in the name of 'popular capitalism'). Unsurprisingly, it was a rightwards political trajectory from which emerged the Blair government ('New Labour') that ensured the continuation of the neoliberal project.

15   Not the least of the many ironies is that a volume (Skinner, 1985) welcoming the return to the social sciences of grand theory featured the ideas not just of postmodernists and their fellow travellers (Derrida, Foucault) but also of others (Lévi-Strauss), all of whom were then in one way or another engaged in challenging both the efficacy of grand theory and its related concepts of modernity-as-systemic-development. At the less serious end of the anti-Marxist spectrum, where spluttering diatribe merges with silliness, can be found the following sort of observation (Putzel, 2004): 'Traditionally, Marxist analysts and activists attempted to explain this reality with appeals to the concept of "false consciousness" [which] has always been one of the more far-fetched propositions of

much current mainstream discussion about development tends to be empiricist, focussing on a specific issue regarded as problematic, solutions to which entail no more than piecemeal change in or amelioration of what exists at present.[16] No challenge is made to capitalism as a system, therefore, and consequently socialism is not considered as a desirable/feasible prospect. Since in this discourse about development-as-modernity capitalism does not feature as a problem, how to go beyond it in terms of theory and practice does not arise, and consequently is not addressed.

Variants on this position include two other arguments, each of which is well represented in the social sciences and development literature. First, the view we don't know how to replace capitalism because we no longer know what it is, or because it is too entrenched or powerful to be challenged. And second, as capitalism is an ever-present system historically, there is no reason why it should – or indeed could – be transcended: because workers are not revolutionary, so the argument goes, systemic development is off the political agenda.[17] Hence, in systemic terms the road ahead differs little from that already travelled, an interpretation shared not just with mainstream views about development, but also with further positions, this time ones that are opposed to concepts of development.

Apart from those who see continued development along the present trajectory as unproblematic or desirable, therefore, other approaches oppose accumulation without, however, wishing to transcend it systemically. Of these, three politically similar kinds of anti-capitalism feature prominently in studies

---

Marxist theorists' This kind of dismissal-by-assertion, unsupported either by evidence or by research, emanates all too frequently from conservative academics who advocate piecemeal, micro-level 'solutions' to the problems wrought by capitalist development. Privileging an unexamined voice-from-below, such arguments arrive at the following sort of conclusion: rather than a case in which I tell you that development is not appropriate for someone like you, better by far ideologically and politically is when I reveal that it is you yourself who rejects this process, which is where the role of academia becomes important. Where the study of development is concerned, therefore, it is difficult not to see this role as negative, akin to the dissemination of false consciousness, in the dual form of combining distraction and misinformation.

16   Possibly the most succinct criticism of privileging empiricism, in the process downgrading or dispensing with theory, remains that made by Thorner and Thorner (1962: 187) about the flawed methodology used by the First Agricultural Labour Enquiry in India, to the effect that 'if it is not clear what constitutes an attached labourer it does not matter whether [one] says that in a given state there are 1.3 per cent or 78 per cent of them'.

17   As is outlined below in Chapter 5 of this volume, these two perceptions are held, respectively, by Jan Lucassen and Marcel van der Linden, both of whom are advocates of what they term global labour history.

of development. The first is that of the conservative/reactionary/populist right, the hostility to capitalism of which takes the form of advocating a return to varieties of pre-capitalist social and economic organization. Opposed to what is regarded as the erosion by development of traditional culture, religion, and hierarchy, it is nevertheless as opposed to socialism for much the same kinds of reason.

A second version argues simply for a return to patterns of accumulation that preceded neoliberalism, on the grounds that – unlike the latter – the former is perceived as being a more benign sort of accumulation project. Such optimism is itself shared with the third version of anti-capitalism, an approach that consists of the 'new' populist postmodernism; its exponents are antagonistic towards development and modernity (seen as harbingers of capitalism), categorized by them as inappropriate Eurocentric impositions on rural populations – especially peasants – in Third World nations.[18] Like that of conservatives, this particular view is as hostile to socialism as it is to capitalism; analytically it re-essentializes the identity politics associated historically with notions of 'peasantness' as innate, unchanging, and unchangeable.

Post-development?

Early critiques of populism, not just by Marxists, have now been replaced with endorsements of this anti-development approach.[19] In an observation that mirrors the Leninist critique and has current relevance, Draper emphasized the political difference and incommensurability between populism and socialism, pointing out accurately that historically populism was 'a peculiar

---

18   There are many reasons why the 'new' populist postmodernism has succeeded in taking the social sciences by storm, but one that deserves mentioning is its seemingly – albeit undeserved – theoretical freshness. This aspect – the idea of theory as fashion – is well captured by Houellebecq (2022: 128) when he observes: 'Unlike their predecessors, the new progressives do not identify progress by its intrinsic content, but by its novelty … they live in a sort of permanent epiphany … where everything that appears is good by the simple fact of its appearance'. All too often new things that haven't been seen before turn out to have been 'missed' for the simple reason that they are not there to be seen in the first place. For more on this issue see Chapter 3 in this volume.

19   As an overview (Walters, 1960: 223) of the debate about populism in America indicated, 'populism had its roots deep in that distinct social entity, the American farmer, who could not become a proletarian, and who, when his way of life was seriously invaded, turned to familiar remedies – managed currency, control of monopolies, and land legislation', concluding that 'populism was the natural expression of farmer protest; socialism was the natural expression of the dissent of industrial labor'.

American device to defend the capitalism of the many against the capitalism of the few', amounting to no more than 'a dream of recapturing an imaginary idyllic past of independent freeholders'.[20] Others at that conjuncture echoed this assessment, underlining the extent to which agrarian populism in America was xenophobic and antisemitic.[21]

On the relationship between culture and development, again the contrast is marked. For postmodern theory grassroots culture is already-existing, a series of customs, traditions, and way of life recuperated from the past, where are found discourses and institutions deemed 'eternal' and 'authentic', merely to be rescued, protected, and once more practiced.[22] By contrast, because Marxism regards grassroots culture in pre-capitalist and capitalist modes as unduly penetrated by their respective upper-class ideologies, an authentically grassroots variant has to be constructed, not reconstituted, and for this reason lies in the future, not the past. Epitomizing this retreat from development are concepts like 'de-growth' or 'post-development'; the litany of processes to which such theory objects is identical to that of the 'new' populist postmodernism, of which it is in effect a variant.[23] For those in the latter categories, emancipatory agency (by what are claimed to be 'new' social movements) involve nothing more than quotidian resistance, the object being either a return to a non-capitalist social order or the realization of a benign (= 'kinder'/'nicer') form of accumulation.

Opposed to 'the core features of the development discourse – economic growth, productivism, the rhetoric of progress, instrumental rationality', post-development is as its name suggests hostile to all manifestations of growth, an antagonistic position which very clearly extends to Marxist theory.[24] Under

---

20   Draper (1957: 36–39).
21   On this see Walters (1960: 224).
22   It is difficult to improve on the wonderfully acerbic definition of just such an approach by Curtis (2015: 1229) who, in the course of reviewing a large *Handbook of Sociology*, observes that 'my experience of cultural studies was that it deals with trivialities, and seems to spend more and more energy arguing about the creative capacities of people across ever narrower domains'.
23   Many exponents of post-development are either agrarian populists or postmodernists, or both. Among those contributing to its programmatic text (Kothari, Salah, Escobar, Demaria, and Acosta, 2019), therefore, are found 'new' populist postmodernists like Arturo Escobar, Vandana Shiva, Gustavo Esteva, and Serge Latouche. Arguments promoting the currently fashionable term 'de-growth' are found in Soper (2020), Jackson (2021), and Hickel (2022), among others.
24   Along with Marxism, Lenin and socialism are condemned in the following dismissive manner (Kothari, Salah, Escobar, Demaria, and Acosta, 2019: 99, 218, 255): as '[s]tructures of domination justified as requirements for the transition to socialism'; as 'two hundred years of environmentally irresponsible growth and development brought about by

the rubric of 'developmentalism' is located every form of advance, from industrialisation, via technology, massification and urbanisation, to modernity itself, forms of negativity attributed to the effects of capitalism, socialism, and western imperial power.[25] Proclaiming the development era to be at an end, post-development theory claims to have emerged from 'indigenous, peasant and pastoral communities', contexts that also form its blueprint for the future. Hence the lament that 'culture, ethics, and spirituality [have been] sidelined and made subservient to economic forces', together with its endorsement of the agrarian populism of Gandhi.

This Sense of Identity

Symptomatic of the influence exerted by populism over theory about development is the dominance exercised in the recent past and currently by the subaltern studies project. Throughout the social sciences (and elsewhere) the term 'subaltern' is now so ubiquitous that it is easy to forget how it was Ranajit Guha who first brought Gramsci's concept to the attention of development studies, not just in India. Opposing the hitherto dominant historiography of India as doubly elitist, imbued with a modernizing and a colonial bias, he advocated the investigation of those below – mainly rural populations in the subcontinent – in order to give a voice to the silenced by reinterpreting sympathetically what had been written about them in official documents and other sources. Central to such an approach would be the construction of a new understanding as to the causes and purposes of agrarian movements during the colonial era, an objective to be realized by the close study of those participating in these sorts of mobilization, from which might be elicited the peasant voice of those involved.[26]

---

capitalism and socialism'; while 'capitalism and socialism prove to be two sides of the same coin'.

25   Kothari, Salah, Escobar, Demaria, and Acosta (2019: XXIV) maintain that '[t]here is a fuzzy line when it comes to productivism, modernization, and progress', because '[w]hat these European Left intellectuals did not anticipate was how, today, alternatives are also emerging from the political margins – from both the colonial periphery and the domestic periphery of capitalism'. This is quite simply wrong, since from at least the mid-nineteenth century onwards Marxists have been only too aware of populist opposition to any form of development (= 'alternatives'), on the spurious grounds that leftist theory about the desirability/feasibility of modernity/progress constituted an inappropriate/alien (= European) project imposed on the colonial periphery.

26   See Guha (1982-89; 1983), whose methods, theory, and the political implications of his analyses were themselves the subject of intense debate over the following decades, mainly

As easily forgotten is the epistemological influence on Guha of his political background in the Communist Party of India (CPI), which adhered to the concept of a 'progressive' national bourgeoisie, that in alliance with undifferentiated peasants and workers would – it claimed – lead to a benign form of development ushered in politically by a democratic stage.[27] Hence the target of CPI politics was not capitalism but feudalism, the latter associated with the landlord class and British colonialism; equally, its political support was not for socialism but rather for a nationalist inclined rich peasantry which, the CPI insisted, would bring about a transition to an as-yet absent process of domestic economic growth.[28] This was the political analysis from which his ideas about an alternative and authentic 'from below' national identity emerged, and as such was instrumental in the way the subaltern studies project was formulated.

Accordingly, for Guha the important divide was the same as that identified by the CPI, between peasants and the landlord class, the latter seen both as allies of an external power, colonial Britain, and as 'feudals'/'semi-feudals' simultaneously exploiters of petty commodity producers. Hence the central role of peasant insurgency at the heart of Indian national resistance, both to late 18th and 19th century colonialism and to the modernizing Nehruvian state, characterized by Guha as elite nationalism. It was in opposition to this duality that the subaltern studies emerged and consolidated, pitching the autonomy of peasant consciousness against both elites, not just external colonialism. His antagonism towards the concept of development is evident from opposition to its application in Uttar Pradesh, described by Guha as 'colonial petty-bourgeois day-dreams that served as a foil to the imperial illusion of grandeur [and] the convergence of both … on a somewhat magical belief in science'.[29]

---

in the pre-2009 *Journal of Peasant Studies* (Arnold, 1984; Bayly, 1988; Sathyamurthy, 1990; Brass, 1991, 2000a, 2002; Hardiman, 1995; Kumar, 2000; Singh, 2002; Beverley, 2004). It is not necessary to agree with him to recognize the fundamental way Guha and the subaltern studies project has influenced academic theory about development and the historiography of agrarian society.

27 On this background, see Amin and Bhadra (1994) and Guha (2010: 8–9). The importance to Guha of his own national and cultural heritage is evident from an observation by Mechthild Guha (2014: 49), his Austrian-born spouse, that 'Ranajit and I discussed many times what our roots were. He feels rooted in Bengali language and culture irrespective of where he lives … believing it crucial to maintain this sense of identity'.

28 Hence the kind of defence mounted by the CPI (Sinha, 1982) of its support for remunerative prices for agricultural produce, a view criticized by other leftist groups for being pro-kulak. Against this the CPI claimed that 'it is not the rich peasants (kulaks) but the landlords who are still the main enemies of the democratic movement', not only confirming its backing for rich peasants but also resorting to the old canard – long disproved – that remunerative prices would enable them to pay their agricultural labour minimum wages.

29 Guha (2010: 119–123).

INTRODUCTION   13

Dismissing accounts of Indian nationalism which locate it at only at the level of the 'indigenous elite' ranged against colonial power, Guha maintained such an approach ignores 'the sturdy nationalism' of the Indian peasantry.[30] This suggests that for him this 'from below' variant was more authentic and benign, part of the peasant-as-the-backbone-of-the-nation discourse, which in turn underpins agrarian populism.[31] The latter was a backwards-looking ideology, seeking to oppose colonialism not from a progressive view but rather from hostility to modernity as an alien, non-Indian idea linked to colonial power. Consequently, peasant movements categorized as 'primordial', espousing 'traditional' ideologies, and composed of subaltern elements, are all ranged against this dual elite of an internal modernising state and external colonialism.[32] This form of populist mobilisation, driven by an ideology consisting of non-modern/anti-modern views, is for Guha the real struggle.[33] Insofar as it took issue an external colonialism, this was consistent with the position held by the CPI, the 'new' populist postmodernism, and the subaltern studies project, for each of which the heroic subject of history was a nationalist peasantry undifferentiated by class.

The frequent dismissal of development – not least by Guha and other exponents of the subaltern studies approach – as a specifically Eurocentric project, and thus not applicable to the Third World, is problematic. To begin with, the

---

30   Guha (2010: 514) rejects the theory about nationalism of Benedict Anderson and Gallagher and Seal as 'colonialist point[s] of view' which conceptualise nationalism at the level of indigenous elite and colonizer, because such accounts fail 'to acknowledge and explain the sturdy nationalism of the mass of the people, especially the Indian peasantry. It was they, and not the loyalist elite, who alone resisted the Raj (often with arms) during the eighteenth and nineteenth centuries, and sketched out, albeit imperfectibly and in quasi-religious idioms, any alternative to British rule that was not designed specifically to restore landed magnates of the pre-colonial period to power'.

31   This kind of idealization about the peasantry and the countryside is evident when Mechthild Guha (2014: 83, 90) confirms that the pattern and livelihood of agrarian society appealed to them both, for whom farming 'was harmonized with nature' and 'rural ugliness caused by ... mindless modernisation'.

32   Referring to the subaltern studies project, Guha (2010: 514, original emphasis) contends that 'recent work on Indian history ... has established beyond doubt [that] much of this movement *originated* in popular initiatives independently of elite leadership [and] derived from the primordial power relations and traditional ideologies that had little to do with either the culture or the institutions of the Raj. Indeed, the Indian experience shows that nationalism straddled two relatively autonomous but linked domains of politics – an elite domain and a subaltern domain'.

33   Even the peasant nationalism of Eastern European is regarded by Guha (2010: 515–16) as benign and progressive (= popular nationalism), despite the fact that it was the locus of what became reactionary far-right political mobilizations of the 1920s and 1930s.

term Eurocentric is a slippery concept, having at least two different meanings. One refers to studies of non-European phenomena, processes, people, carried out by those based in Europe, and of European nationality, the inference being that because of this it is impossible for them properly to comprehend and interpret what is being examined. Another refers to studies utilizing concepts/theory originating in a European intellectual milieu (philosophy, political economy, ideology) in order to analyse non-European contexts, the inference being that, given their origin, such frameworks are not appropriate for an understanding of what has occurred and is happening in the Third World. Whereas the first meaning applies simply to the fact of being European as a bar to understanding, the second inculpates not the identity of the researcher him/herself, but rather of the tools s/he deploys.

Climate, Class, Risk

Another debate which poses similar difficulties for leftist politics concerns the link between development and climate change. The latter is in effect opposed to development, and like subaltern studies is part of a wider discourse linked to populist concepts like degrowth and post-development, displacing class and capitalism as problems and allocating blame (and solutions) to humanity in general. Since we are all threatened, so this argument goes, we will all act together to solve the problem, a notion that cannot explain the continuing ability of nations and corporations whose economies are based on fossil fuels to water down climate change policies/initiatives, or indeed prevent them from being implemented.

Where climate change is concerned, therefore, environmental interests are said to predominate, overriding all other considerations.[34] Significantly, this kind of view is linked to and creates a space for populism, with its pro-rural, pro-peasant, pro-small-scale ideology combined with opposition to industrialization, urbanization, and socialism. Hence the view that climate and not class struggle and socialist transition should be first on the political agenda. The difficulty with this is that does anyone really think that capitalism is – or

---

34 This is the view, for example, of Žižek (2023: 21–22): 'What needs an explanation is the basic madness of the situation: at a time when it is generally agreed that our very survival is under threat for ecological reasons, and when everything we are doing should be subordinated to cope with this danger, all of a sudden the prime concern has become a new war [in Ukraine] that can only shorten our passage to collective suicide … [t]o explain this through the interests of big capital and state control falls short'.

can ever be – a force for good in formulating policies and ensuring their implementation in order to prevent – let alone reverse – climate change?[35] The only politics and policies capable of achieving this are socialist ones, with their emphasis on large-scale and long-term economic planning, designed to benefit social formations and carried out on an international level, rather than the much narrow focus of individual corporations the sole concern of which is profitability and survival of the enterprise in the context of ever-more acute market competition.

Combatting climate change should be high on a socialist agenda, certainly, but pursued as part of that, involving crucially policies to do with planning and economic development. In short, not apart from such considerations, as an end in itself to be pursued without reference to the wider system in which this change is to take place. Specifically, not in terms of an attempt to 'reform' capitalism, or to persuade (and await on) the arrival of 'enlightened' capitalists who, so the argument goes, might be coaxed into becoming 'nice' employers who recognize wider social interests rather than ones of profit and survival. Leftists who follow the latter path are engaged in a fool's errand, as a result – again – of looking the wrong way, in effect avoiding the gravity of the issue confronted.

Not the least of the many ironies emerging from the way climate change features in the development debate is how it has wrong footed academics who champion agrarian populism.[36] The latter have combined in a single cause both environmental issues and rural producers, claiming these possess a uniform objective, as represented in 'farmer first' narrative allocating rural producers a central role in the conservation and protection of nature. In the vanguard of the fight against both capitalism and climate change, therefore, would be found not the working class but rather an alliance of non-class elements (tribals, NGOs) and peasant farmers all intent on defending nature. Now, however,

---

35 As always, capitalists in the UK have taken advantage of environmental discourse in order to further their interests and increase profits. A case in point is the decision taken by Tata in January 2024 to cut back steel production at its Port Talbot plant, a policy justified in terms of replacing a carbon intensive operation with a less carbon intensive one, the resulting decarbonization enabling it to meet the net zero emission target. However, it transpires the change also means that Tata will be able to import semi-finished steel from India. In short, a process of outsourcing production to India, where labour-power is much cheaper, or green-washing what is in effect a restructuring of the capitalist labour process. See 'Tata to close last Port Talbot blast furnaces', *The Financial Times* (London), 20–21st January 2024.

36 For these exponents of agrarian populism, see Brass (2021b: Chapter 4) and Chapter 4 in this volume.

throughout Europe these same cultivators are mobilizing *against* environmental policy and legislation designed to achieve net zero emissions, on economic grounds (cost of farm inputs is too high, price of farm output is too low).[37]

Unmissable in this connection is the plethora of avoidance mechanisms designed to reassure everyone that capitalism is capable of reform, is essentially benign and can thus be relied upon still to benefit society at large. These extend from nudge economics to effective altruism, theoretical approaches heavily endorsed by corporate managers, whereby capitalists retain both wealth and power but are encouraged merely to contribute donations to 'deserving' social projects and requested more generally to act 'nicely' in the wider public interest.[38] Variations on this 'forget what divides us as we are all in this together' kind of discourse include currently fashionable sociological frameworks informed by concepts like structuration, individualisation, and global risk conflict.[39]

Seemingly the object of such pointlessly crafted tautologies is to repackage more radical forms of opposition to capitalism in order to make them suitable for consumption by bourgeois academia. Central to these approaches is the theoretical downgrading of class and its form of struggle in the name of aporia, instead privileging epistemologies addressing macro- and micro-level identities above and below class: ranging from on the one hand an at-risk humanity to on the other the individual search for solutions to systemic contradiction.[40] Because a broad notion of risk is universalized and operationalized at the level of globalization, the possibility of solutions that ignore or downgrade class distinctions is correspondingly idealized.[41]

---

37 See 'The rise of agricultural populism', *Financial Times* (London), 6th February 2024.
38 For examples of this sort of pro-capitalist theory see Thaler and Sunstein (2008) and MacAskill (2015).
39 Among the purveyors of such corporate balm are Giddens (2003), Hutton and Giddens (2000), and Beck (Beck and Willms, 2004).
40 'Instead of individualism being placed in context and relativized by class analysis', contends Beck (Beck and Willms, 2004: 101), 'in order to understand class we now need to place it in the new context of individualization … [a]nother point that is important for class analysis is that this sort of individualization necessarily entails a decline in overarching social narratives, paradoxically by multiplying them so that no single one can achieve an undisputed hegemony'. In short, we are once again placed in the postmodern realm of fragmentation and undecidability.
41 Hence the politically over-optimistic view (Beck and Willms, 2004: 139) that '[r]isk conflict abolishes existing barriers to communication, and puts people into communication who don't want anything to do with each other'.

Themes

Although this book pursues many of the same themes as an earlier volume, and can thus be seen as a companion piece to the latter, its net extends to include more recent contributions to the development debate, in the process addressing the political direction in which this discussion is going.[42] The first part, consisting of Chapters 1-4, looks at interrelated issues that feature centrally in development studies, extending from ideology about race, together with the question of racism, its occurrence and reoccurrence, to the industrial reserve and labour market competition. Of interest is how these issues are interpreted currently through the lens of the agrarian myth, peasant essentialism, agrarian populism, and identity politics. Why this is so is linked in turn to the the kind of power and influence exercised in academia over prevailing ideas and their dissemination.

The themes examined in the first part of this volume are pursued in more detail in the second part. Thus Chapters 5-8 illustrate the direction in which recent contributions by van der Linden, Fukuyama, Hart, and Breman to the development debate are taking the discussion, together with its political implications and the reason for this. Of the four, it is argued that only Hart can be regarded as having made a positive and enduring contribution to these discussions, albeit in terms of a critique based on the concept informal sector economy. Models associated with the three other approaches, extending from global labour history, an attempted recuperation of classical liberalism, to neo-bondage, are shown to be methodologically and theoretically flawed. The third part consists of Chapters 9-10, each of which considers how postmodern theory has penetrated social sciences generally and development studies in particular via claims by social historians to have found 'authentic' grassroots voices hitherto condemned to silence.[43]

---

42   The earlier volume is Brass (2022b).

43   Although the focus here is on the way development theory has been colonized by postmodernism, it is impossible to over-exaggerate how the influence of the latter has spread across the cultural domain in western capitalism. A case in point is what are termed young British artists, as represented by the work of Damien Hirst, Tracy Emin, and others. This genre, described (Stallabrass, 2006: 21) as a 'facile postmodernism [that] was the foundation of this new art, one which took no principle terribly seriously, which pretended not to separate high from mass culture and which, given this relativism, accepted the system just as it was, and sought only to work within it'. Its avant-gardist claims notwithstanding, the essential conservatism of this movement is difficult to disguise. Forming part of celebrity culture, in which the artist is more important than the art, it was emblematic of Thatcher's Britain, in that it seemed to valorize the cultural aspirations of those

Contrasted in Chapter 1 are the different ways of interpreting ethnic identity, and accounting for racist ideology linked to this. Because notions of cultural otherness hid economic differences, Marx argued that racism would cease only when the economic conditions generating its reproduction also ceased, formulating a materialist interpretation of race and the development process informing leftist theory. By contrast, postmodernism dehistoricizes ethnic/national identity, declaring it an innate, eternal, and thus empowering form of belonging. The latter interpretation is also found earlier, in fictional narratives about the Reconstruction era in the post-bellum American South, as seen by racist whites. For the latter, policing emancipated black ex-slaves and white female sexuality overrides the economic causes of racism, the importance of which was demonstrated by sociologists conducting research into the connection between lynching and labour market competition involving ex-slaves and landless whites in the South during the Great Depression.

Chapter 2 outlines how and why an epistemological and political shift in the meaning of the industrial reserve army, an important concept of political economy generally and Marxism in particular, has taken place from the nineteenth century onwards. For early Marxists, therefore, an effect of the globalization of the labour regime was that, by encouraging competition for jobs, surplus labour could regulate the market on behalf of capital, keeping wages down, and discouraging class solidarity and organization. Although recognizing the deleterious impact on worker unemployment and impoverishment of a growing industrial reserve, liberals such as Beveridge advocated greater state intervention (regulation, social insurance) in order to save capitalism and prevent socialism. Subsequent leftists such as Kalecki, Sweezy, and Dobb also regarded the industrial reserve as negative, warning that because capital opposed full employment, it would always seek access to foreign labour, an objective increasingly made possible by globalization and deregulation. Currently, the latter permits employers either to outsource production to where labour is cheaper, or insource labour itself, allowing them to restructure their workforce. By contrast, much postmodern theory views the industrial reserve as unproblematically benign and positive, arguing that as open-door migration is a 'human right', borders – a legacy of colonialism – should be abolished. Unlike Marxism, which perceives the industrial reserve as disempowering for the existing workforce, a weapon used against the latter by employers, postmodernism regards it as a revolutionary source of anti-capitalist mobilisation.

---

below, indicating that even plebeians could become – as some of them did – wealthy and famous, perpetuating thereby the neoliberal capitalist myth of upward social mobility.

Via academic publishing, the focus of Chapter 3 is on the fictional narrative of *The History Man*, a novel by Malcom Bradbury, which expressed in the domain of popular culture the hostility felt by the academic establishment towards Marxist theory as taught in sociology departments of the new universities. To a large degree, it was this account that constructed the negative public image of the sociologist-as-Marxist, depicting its practitioners as intellectually unrigorous/unprincipled and venal purveyors of political subversion, in the process constructing a series of images (empowered dilettantism, politics as fashion accessory) that delegitimized any/all forms of radical leftist commitment. The irony of discrediting Marxism as a malign academic influence is hard to miss: not only did this diminish, replaced by the anti-Marxist cultural turn, but a number of leftist intellectuals were prevented from entering or remaining within social science departments. This process has also had an impact on what is and what is not produced by academic publishers.

Chapter 4 explores in more detail the way in which populism has become hegemonic in the kind of theory now applied to rural development, together with the reasons for this. Hence the claim by exponents of what is termed Critical Agrarian Studies to have redefined the way peasant economy and culture is understood, in the process replacing the hitherto dominant paradigm based on modernisation. What this new approach entails is nothing less than depriviliging Marxism and repriviliging agrarian populism: the epistemological focus of analysing rural society has shifted, no longer being about the internal differentiation of the peasantry and the disempowerment of class, but rather the empowerment of subsistence-oriented family farming. Advocates of this agrarian populist approach support either a return to a pre-capitalist social order or to a more benign sort of accumulation. Contrary to what is claimed by its adherents, therefore, Critical Agrarian Studies is neither new nor radical: much rather, it is in keeping with the wider post-1980s conservative political shift towards individual private property ownership and so-called 'popular capitalism'.

Turning to more specific kinds of analysis, examined critically in Chapter 5 are two seemingly antithetical views about development. On the basis of what it terms an 'extended working class' that includes an undifferentiated peasantry and the lumpenproletariat, van der Linden holds that capitalism is found everywhere, at all times, and in most places. The other, by Lucassen, expels capitalism, class, and class struggle, from its analysis. Each approach, however, shares a number of problematic characteristics, not least a failure to differentiate politically distinct forms of anti-capitalism. Both are informed by what is claimed to be a 'new' paradigm, global labour history, the epistemology and concepts of which depart from comprehending – and thus constitute a denial of – the systemic process that is capitalist development. The latter, as understood by Marxism and other social science theory, is an historically and – initially – locationally specific mode of production; the resolution of its inherent crises and contradictions also require its systemic transcendence, and replacement by socialism.

Perhaps the most optimistic and relentless academic defence of capitalism as a system of development is still that held by Francis Fukuyama, whose ideas are considered in Chapter 6. It is a defence-by-proxy, since what is deemed in need of protection is classical liberal theory, of which the accumulation process and bourgeois democracy are integral and benign parts. These require safeguarding from a double threat they pose to liberalism, as he sees it, consisting of on the one hand *laissez-faire* economics from its right, and on the other identity politics from its left. Each is regarded by him as an alien excrescence that deforms classical liberalism, as such unconnected with capitalism itself. Ignored thereby is much rather how both neoliberalism and identity politics are effects of capitalist development: the former as the market and its form of competition spreads across the globe, to which the latter politics are the populist response.

Examined in Chapter 7 is the contribution made by the anthropologist Keith Hart to the study of development, and why this is important. Despite personal travails (health issues, peripatetic academic employment), a revealing and self-reflexive autobiography by him traces the emergence of the informal sector concept to participant/observation fieldwork he conducted in Ghana. Along with a few others at the time, Hart challenged the dominant modernisation paradigm, which argued that societies in the Third World would replicate the development path followed by metropolitan capitalist nations. Instead, and on the basis of research findings in Accra, he outlined how inhabitants of urban slums were neither marginal nor fatalistic but socially dynamic as a result of being involved in economic activity, an interpretation that has since become influential in development studies. Other contributions by him to the latter include promoting the idea of an anthropology practiced outside the formal boundaries imposed by academia.

Compared in Chapter 8 are two paradigms addressing the same issue: the acceptability to capitalism of labour-power that is unfree, a debate about the trajectory followed by the capitalist labour regime in which geographers – particularly those dealing with development issues – are now taking an interest. Conceptualized either as deproletarianisation or neo-bondage, the definitions of these two approaches, together with the purpose of the production relation involved, overlap to a large degree. Traced, therefore, is the emergence, the theoretical consistency, and the political implications in terms of systemic transition of each concept. Although epistemologically indistinguishable with regard to issues like employment duration, the kind of worker affected, and enforcement from within the kinship domain, a political difference exists: whereas neo-bondage envisages a resolution involving the return to a 'caring'/'kinder' capitalism, deproletarianisation by contrast requires a transition to socialism.

Finally, Chapters 9 and 10 consider both the fact of and the reasons for the vanishing from academia in general, and the social sciences in particular, of critique based on Marxist theory. Each shows how social history has played a major part in licencing the 'new' populist postmodern colonization of the social sciences, a process facilitated by the claim to have uncovered what the voice-from-below really thinks and does. Outlined in Chapter 9, therefore, is how just such a paradigm – applied by Patrick Joyce initially to the analysis of the urban working class in Victorian England, advocating the replacement of concepts like modernity, grand narratives, and social determination with alterity, difference, and aporia – has now been applied by him, unsuccessfully, to the peasantry. Although his attempt to uncover an 'authentic' rural smallholding proprietorial voice endorses populism together with Chayanovian theory, privileges folklore as a source, and rejects Marxism, he nevertheless unwittingly and contradictorily smuggles in a Leninist model of peasant differentiation, albeit decoupled from a socialist transition.

Chapter 10 looks at how, despite appearances to the contrary, Marxism appears under siege in journals and books, its concepts and framework deemed irrelevant or Eurocentric when applied to explanations of present-day accumulation, as such challenged or discarded even by publications ostensibly sympathetic to its aims. Marxist analysis is increasingly replaced by supposedly fresh paradigms, a change driven in part by academic competition, involving a search for relevance, popularity, and funding. Such non- or anti-Marxist alternatives, often formulated by ex-Marxists (Laclau, Mouffe), deprivilege class, revolution, and socialism, advocating instead a 'new' populist postmodern approach privileging discourse about national/ethnic politics. This is justified by postmodern theory on the grounds that, as these are innate kinds of identity, such authentic ideologies are consequently more progressive and empowering for those at the grassroots. It is argued that this epistemological shift can be traced, again in part, to the post-1960s entry into academic posts of Marxists, many of whom transferred their political allegiance to the 'new' populist postmodernism, misinterpreting hegemony exercised by 'the people' as unmediated progressive voice.

PART 1

Questioning the Paradigm

∴

CHAPTER 1

Racism and Development: Blood, Sweat and Fears

> The first fundamental rule of historical science and research, when by these is sought a knowledge of the general destinies of mankind, is to keep these, and every object connected with them, steadily in view, without losing ourselves in the details of special inquiries and particular facts, for the multitude and variety of these subjects is absolutely boundless; and on the ocean of historical science the main subject easily vanishes from the eye.
> Historiography as it should be practiced, according to FREDERICK VON SCHLEGEL.[1]

∴

Introduction: More Lessons from History

Impossible to miss nowadays is the presence of a close link between three apparently disconnected phenomena: ideas about the nature of development, questions of class and non-class identity, and the process of labour market competition. As unavoidable is the current divergence between Marxist and postmodern approaches to this connection, not least in terms of radically different interpretations of the role played by the privileging of ethnic or national 'otherness' in economic situations where the search by labour for employment is acute, together with the kind of ideology and struggle licensed thereby. Whereas historically an ability to conduct struggle 'from above' – one favouring capital over labour – has been restricted to particular national contexts, with the onset of development, and a concomitant decrease in skill levels required of workers coupled with an increase in the source and quantity of the industrial reserve in a context where capitalism itself is deregulated, employers are now able to access labour world-wide. In such circumstances, racism can and does arise where capital draws on an enhanced reserve army, generating acute competition for jobs between workers of different ethnic/national identity.

---

1 von Schlegel (1848: 69).

This sort of rivalry is itself fostered by employers for two reasons: to maintain or enhance profitability when competing with other producers in the market; and in order to pre-empt or prevent the emergence or consolidation of consciousness based on class, a solidarity which might threaten the ownership/control of the means of production/distribution/exchange currently enjoyed by capital. How a linkage involving these seemingly disparate phenomena can play out historically is illustrated here with regard to the United States over the late nineteenth and early twentieth centuries, when each of these processes was evident in the way racist ideology was fostered and reproduced – both from above and also from below – in the post-bellum South, both in the Reconstruction era and also in the Great Depression.[2]

Ideas plus the conflict to which they gave rise in that context at that conjuncture anticipate, and thus have lessons for, current debates about development paths to be followed, and their desirability or undesirability. Highlighted in particular is the difference between two distinct approaches to development: between one which maintained that the most effective struggle against capitalism is (as Marxists argued) based on class and revolutionary agency designed to bring about socialism, while the other (as postmodernists maintain) is based on non-class identities – including ethnicity and nationality. Each addressed the same political and ideological effects of the same economic developments – among them the rise in Western capitalist nations of racism/populism/nationalism – but with a radically different set of assumptions, concepts, and outcomes.

Since it is a link that in the past has been recognized analytically by Marxist and non-Marxist alike, it would not usually be an epistemological requirement to have to underline once more the enduring connection between on the one hand racism, nationalism, and populism, the form and intensity of their reproduction, and on the other labour market competition. Many leftists, and all socialists extending from Marx and Engels, via Lenin and Trotsky, to Deutscher and Glyn, have emphasized its centrality to capitalism, as have non-Marxists, among them Beveridge and Sartre.[3] That it is necessary to do so again

---

2   Despite the fact that in the historical texts analysed here the language of race is today considered inappropriate, it has been retained when citations are made, not least to underline the element of racist ideology informing them. It must be emphasized, therefore, that to cite the language of the time in this way is not an endorsement of what is said, much rather the opposite. For the genealogy of the concept 'race', see Augstein (1996), while the background to its deployment and reproduction in the United States is outlined by Newby (1968).

3   For their views linking the reproduction and intensity of nationalism, racism, and populism to labour market competition, and the latter in turn to market rivalry between capitalist producers, see Brass (2022a: Chapter 7; 2022b: Chapters 1 and 2).

can be traced to the hegemonic way in which race and racism are currently interpreted by exponents of postmodernism, for whom non-class identity is traditional, innate, and unchanging, a series of characteristics applied by them both to victim and perpetrator of racist ideology.[4] Unsurprisingly, therefore, decoupling the recuperation of racist ideology and the economic intensification of labour market competition ignores what history teaches about the significance and effect of this articulation.[5]

This chapter is divided into four sections, the first of which examines the views on race held by Marx, who linked both racism and its eradication to transforming the material conditions which generated and reproduced this ideology. That postmodernism, by contrast, declares both non-class identity and hostility to it as innate, is considered in the second section. The third looks at the way race is depicted in fictional accounts of the American South during the Reconstruction era, while the fourth compares this with the findings of sociologists conducting research there during the Great Depression.

I

An Absent Nationality

Clearly, the focus of Marx when considering the question of race is on Judaism, not on blacks. Nevertheless, it permits one to understand his approach to the issue of ethnic 'otherness' in general, the specificity of its case study notwithstanding. In contrast to interpretations that categorize both race and racism as innate, Marx and Engels argued that behind the apparent specificity of cultural 'otherness' – bolstered in the case of Judaism by religious practice – was an economic position shared with those who did not belong to the same culturally-defined identity.[6] The same is true of the way in which hostility

---

4 On this see Brass (2021b: Chapter 7). Unsurprisingly, perhaps, this in turn has generated analyses (Hochschild, 2016; Isenberg, 2017; Williams, 2017) in which not only is victimhood linked to 'otherness' reallocated to a different category, but labour market competition together with the advantages to capitalists of a divide-and-rule strategy are either unmentioned or only of marginal significance.

5 As many instances throughout history confirm, an important reason why on occasion race becomes a political issue is when its role as a divide-and-rule tactic is used by capitalists in the class struggle. This emerges in situations where employers in conflict with a workforce composed of different ethnicities favour one identity over another, with the object of undermining any class solidarity achieved hitherto.

6 'Once Jewry was stripped bare of the *religious* shell and its empirical, worldly, practical kernal was revealed', argued Marx and Engels (1975: 108–9, original emphasis), 'the practical, *really social* way in which this kernel is to be abolished could be indicated. Herr Bauer was

separating those possessing different racial identities in the American south during the 1930s hid the fact that many of the antagonists nevertheless shared a common economic position (see below). In that context, at that conjuncture, racism flourished because those with different identities were also rivals in the search for jobs and tenancies, a process of labour market competition which accumulation generated and on which it depended.[7]

The view held by Marx about Judaism, and indirectly about race, were formulated as a response to the case made during 1843 in *Die Judenfrage* by Bruno Bauer. Questioning the exceptionalism of the Jews, who wanted political emancipation that recognized their separate identity within Germany, Bauer argued that why should they – as Germans – advance a claim to a different identity within the nation, instead of fighting for the political emancipation of all – non-Jews as well as Jews.[8] Marx approved of the way in which Bauer posed the question, but disagreed with his solution, which was framed solely in terms of religious belief: the position of Judaism in a Christian Germany. Rather than a theological issue, argued Marx, the question was – and could only be posed as – a political one. Of additional concern to him, moreover, was that once political emancipation had been achieved, the leadership of the Jewish community threw in their lot with the counter-revolution.[9]

Where Judaism is concerned, therefore, Marx indicates that he wants to move beyond its formulation by Bruno Bauer, simply as an issue of religious 'otherness', and focus instead on the link between emancipation and 'what particular social element has to be overcome in order to abolish Judaism'. To this end, Marx proposed to examine not the religious 'otherness' of being Jewish, but rather 'the secular basis of Judaism' – its material foundation. This, he argues, consists of all the economic tropes ('huckstering', 'money', finance) which fuel anti-semitism; it is from the latter (economic + ideology) that

---

content with a "religious question" being a "religious question" ... Consequently Herr Bauer has no inkling that the real *secular* Jewry, and hence *religious* Jewry *too*, is being continually produced by the *present-day civil life* ... Jewry has maintained itself and developed *through* history, *in* and *with* history, and that this development is to be perceived not by the eye of the theologian ... not in *religious theory*, but only in *commercial* and *industrial practice*'.

7 Aspects of the link between racism and labour market competition, including how this connection is interpreted by other Marxists (Lenin, Trotsky, Deutscher) and non-/anti-Marxists (Gobineau, Wagner, Lawrence), are also considered elsewhere (Brass, 2017b: Chapter 19; 2022a: and 2022b: Chapters 1 and 2) and in Chapter 3 of this volume.
8 On the case made by Bauer, see Marx (1975: 146).
9 See the observation by Marx (1977: 32, original emphasis): 'And as for the Jews, who since the emancipation of their sect have everywhere put themselves, at least in the person of their eminent representatives, at the *head of the counter-revolution*'.

'the self-emancipation of our time' must be effected. Here is the crux of the racism issue as seen by Marx: the negative image of Jewishness, and with it anti-semitism, will cease only when the economic conditions fuelling them – capitalism as a system – are transcended. When this happens, the religious element structuring Jewish 'otherness' would also vanish, 'in the real, vital air of society' – one that was no longer capitalist.[10]

Marx insisted that Jewish emancipation was premised not on the separation of Judaism from the state, but rather on the decoupling of the state from religion in general. Hence the focus of Marx is not religion but the state, and its development towards a secular institution. In short, he reverses the usual way of posing this question: instead of the desired path being a recognition of the particularistic (religion, Judaism), it moves in the opposite direction, onto the terrain of the general, as embodied in a state disconnected from *all* religion.[11] The power of the Christian state, argued Marx, resides in its religion, whereas the democratic state, by contrast, does not need religion, since the desires/aspirations embodied in religious belief are met in a secular fashion – a materialist interpretation of the development process. This, of course, is contrary to the kind of emancipation advocated nowadays by exponents of the 'new' populist postmodernism, who not only privilege cultural 'otherness' but demand that such identity be recognized and enshrined legally by the state.

For this reason, he contended, it is necessary for Jews to work not for their own emancipation but rather for 'human emancipation', for others as well as for themselves. In this way, they will negate their present condition of alienation, by realizing a selfhood that was not different (= being 'other'). Rather than seeking emancipation as Jews and no more, which according to Marx will not solve the problem of estrangement (= the identity of 'otherness'), his view

---

10   Noting that '[w]e are trying to break with the theological question', Marx (1975: 169–70, original emphasis) elaborates: 'For us, the question of the Jews's capacity for emancipation becomes the question: what particular *social* element has to be overcome in order to abolish Judaism? ... Let us not look for the secret of the Jew in his religion, but let us look for the secret of his religion in the real Jew. What is the secular basis of Judaism? *Practical* need, *self*-interest .... Emancipation from *huckstering* and *money*, consequently from practical, real Judaism, would be the self-emancipation of our time. An organization of society which would abolish the the preconditions for huckstering, would make the Jew impossible. His religious consciousness would be dissipated like a thin haze in the real, vital air of society'.

11   'We do not turn secular questions into theological questions', argued Marx (1975: 151, original emphasis), 'We turn theological questions into secular ones. History has long enough been merged in superstition, we now merge superstition in history. The question of the relation of political emancipation to religion becomes for us the question of the relation of the political emancipation to human emancipation'.

is that necessarily 'the emancipation of the Jews is the emancipation of mankind from Judaism'. Ceasing to be – and to be seen as – socially 'other', both in cultural and religious terms, and with it the racist stereotypes informing antisemitism, depends in turn on a corresponding willingness to disengage from capitalism itself: that is, to address the prefiguring economic conditions reproducing the ideology about cultural/ethnic 'otherness', both for its subject and for those not belonging to this category.

Using the example of Jews in Hungary during the 1848 revolution, Engels is critical of the tendency of minorities to invoke an ethnic 'other' cultural identity within the larger national unit, and then on this basis to claim what he regards as special treatment (= particular privileges).[12] This is a critique that applies just as much to the attempt currently by the 'new' populist postmodernism to do the same nowadays on the grounds of possessing another – non-class – identity. It is also consistent with the argument made six decades later by Lenin: to oppose strongly the repression of minorities, *but* not to promote the 'other' identity of those same elements. This distinction is crucial, in that it separates two distinct ideological processes: one is a struggle against – with the object of preventing – the economic and/or political subordination/oppression of an ethnically or nationally 'other' minority group; the other, however, is merely to endorse 'difference' and then to mobilize solely because of this.

II

A Race against Time, a Time against Race

Rather than looking at how the 'otherness' of race changes over time, and why, postmodernism approaches the issue by essentializing both race (as an identity) and racism (as the perception of that identity). Dematerializing history in this manner has consequences, not just epistemological ones. Without a dynamic process of development-as-modernity, therefore, it becomes difficult – if not impossible – for this discourse to be a source for the provision of real-life lessons on which to build a world-view and its associated political movement. Accordingly, postmodern theory is a-historical in the sense that it fails to place race and racism in a dynamic context, simply locating the cause of these phenomena in a long-past epoch unremembered by those accused of

---

12 Engels (1977: 232) writes: 'But the Hungarian Germans, although they retained the German language, became genuine Hungarians in disposition, character and customs. Only the newly introduced peasant colonists, the Jews and Saxons in Transylvania, are an exception and stubbornly retain an absent nationality in the midst of a foreign land'.

subscribing to this ideology.¹³ Both the cause and the reproduction of negative images about a particular ethnic identity are decoupled from the conjuncture in which they appear, preventing the situation of racist ideology and its determination in the more immediate process of labour market competition.

Discourse promoting the empowerment of one ethnic/national identity, because of its disempowerment hitherto, is currently not merely accompanied by but also entails proclaiming simultaneously the reverse process: the disempowerment of another such identity, hitherto empowered, whereby the latter is dismissed as a manifestation of an all-encompassing racism.¹⁴ Hence the frequent invocation of the terms 'white privilege' and 'toxic whiteness', undifferentiated concepts about ethnic/national identity applied uniformly to persons/categories across a wide social range, from individuals, via institutions, to whole populations, any/all of whom are labelled innately/irredeemably racist. In the case of present-day Britain, racism – particularly that found among those belonging to the working class – is interpreted by exponents of postmodernism simply as a long-standing and ineradicable effect of nostalgia for Empire and its form of colonial rule.¹⁵ This despite the fact that contemporary holders of such racist views themselves cannot be said either to have had direct experience or indeed knowledge of that nineteenth century imperial/colonial era and its historical processes.¹⁶

---

13 For a symptomatic claim about the all-embracing, fixity, and undifferentiated nature of views about race ('the hitherto unconquerable might and right of whiteness'), see the introduction by Alibhai-Brown to the volume by Bhopal (2018: XIII–XIV), where one encounters the following: 'Bequeathed advantages are found in every nation, among almost all ethnic and racial types ... from the age of exploration, when Europeans set off in search of new lands and profits ... white privilege has been dominant across the globe. Furthermore, through the centuries, relentless Caucasian expansionism and hubris persuaded large numbers of non-white peoples of their own unworthiness. You see "native" humbleness in almost every developing country. Aid agencies, tourist companies, big Western businesses and Christian missions in the 21st century all perpetuate the pernicious notion that white men and women are more evolved and of a higher order than the rest of humanity'.

14 These themes can be discerned in, for example, Tharoor (2017) and Mishra (2020).

15 When presented, evidence in support of this contention is problematic. Thus a YouGov poll is invoked (Sanghera, 2021: 186–87) to support the view about widespread popular support for Empire, notwithstanding a year on year decline in the figure itself plus a failure to interrogate the questionnaire. Commentators endorsing Empire are similarly noted, without however mentioning their conservative politics. As problematic methodologically is the attempt by Kundnani (2023: 154ff.) to explain the meaning of Brexit by confining his interpretation to how it was understood by the UK ethnic minority population.

16 Unsurprisingly, perhaps, when considering the global impact of imperialism, Sanghera (2024: Chapter 3) also misinterprets indenture in nineteenth century British colonies. Failing to mention the circulation in the free market of labour-power that is unfree, he

A variant of this position maintains that the ethnic minority population in the UK opposed Brexit not because they objected to immigration per se but rather on the grounds that it privileged white as distinct from non-white migration.[17] Hence the view that such opposition stemmed from the perception of the EU as a '"white fortress" that facilitated immigration [from other EU countries to the UK] while obstructing the entry of non-white people', an interpretation bolstered by the fact that '[t]he end of mass immigration from the Commonwealth coincided with British accession to the EC [European Community]'.[18] This, it is argued, 'complicates the conventional narrative of Brexit in relation to immigration, which has been shaped by the tendency to see it as an example of "populism"'.[19] Consequently, this interpretation suggests, the UK withdrawal from the EU should be regarded as 'a kind of rebalancing of the UK's focus away from Europe … towards the rest of the world and especially the Commonwealth'.[20]

In short, not an end to labour market competition so much as its redirection, a position amounting to an ethnically-specific post-Brexit open-door policy of engagement with erstwhile colonies that its author thinks the British left should follow.[21] Privileging an expanding industrial reserve in this manner, by presenting it merely as a positive 'rebalancing', chimes with the way postmodernism sees identity as a question simply of culture, not political economy. What this sort of approach overlooks, therefore, is the role culture discharges as a proxy for economic rivalry, perceived by those who invoke the 'otherness' of a contender for employment as a way of protesting against undermining a

---

  instead relies on revisionist accounts by neoclassical economists (Lal, Shlomowitz) which incorrectly depict such production relations as empowering for their subject.

17 Although rejecting the argument that Brexit was simply an expression of white colonial/imperial nostalgia, Kundnani (2023: 156, 170–71) then seems to accept that, after all, the idea of empire did indeed have an influence on Brexit.

18 Kundnani (2023: 160ff., 166).

19 Kundnani (2023: 153, 154, 157–58) challenges equating Brexit with 'white anger', a view he regards as 'centrist' because it is based on the concept populism.

20 Kundnani (2023: 10–11, 168).

21 Endorsing the now-familiar immigration-as-reparations argument, therefore, Kundnani (2023: 171, 178) advocates a 'rebalancing [of] the way the national [UK] story is imagined away from an exclusive focus on Europe', and continues: 'Such a rethinking of British identity and history would have extensive policy implications – especially for foreign policy. In particular, the UK would seek to develop closer relationships with its former colonies [and a] good place to start would be immigration policy, the focus of which has shifted away from the Commonwealth to Europe … It would be possible to go further in the rebalancing of British immigration policy that has taken place since Brexit – in particular by making it easier for citizens of Britain's former colonies to come to the UK'.

standard of living they have enjoyed hitherto, a threat to working class livelihoods implied by an increase in labour market competition.[22]

Where race and racism are concerned, therefore, postmodernism involves the negation of history-as-development in two distinct ways: decoupling narratives about the past from the present, and also from determination by political economy. As significant in terms of development is that postmodern 'experimental history' soon overlaps with the holy grail of populism: the discovery (yet again) of an unheard 'other' voice 'from below' which, it is claimed (yet again) has been silenced by Marxism, modernity, and their respective historiographies.[23] The antithesis of universalism is the postmodern claim that traditional culture as reproduced at the rural grassroots is pristine, ancient, and for its subjects 'natural' and thus largely unquestioned. Regardless of change in the wider economy of which they are a part, rural communities are perceived as outside and against history, always and everywhere resisting the erosion of their culture.

Against this postmodern view, the infrastructure/superstructure interpretation of Marxism argues that historically the pattern of culture has in most contexts been dependent on growth in the economy. Hence the division of labour associated with more advanced forms of economic development permit a society to allocate a portion of its workforce to undertake creative tasks required by the production of art, music, literature, and more generally knowledge. Insofar as capitalism universalises economic processes and institutions, so too does it tend to contribute towards the spread of a broad cultural uniformity. This, of course, is not absolute, but the general trend is unmistakeable: superstructural formation without any reference to the overarching mode of production is historically difficult.

---

22 Interpreting opposition to migration as a cultural threat to the European way of life, as does Kundnani (2023: 6) in his conceptualisation of 'Eurowhiteness', avoids the issue of the historical link between racism, populism, and labour market competition, and how these aspects are all interconnected. This linkage is explored below, in this chapter and also in Chapter 2.

23 See, for example, the following disclosure by an exponent of postmodernism (Munslow and Rosenstone, 2004: 54): 'I discovered that I wanted to write the history of the "other side of the bearch", of indigenous island peoples with whom I had no cultural bond, of Natives. And on "this side of the beach", my side as an outsider, as Stranger, I wanted to write the history of people whom the world would esteem as "little". I wanted to write history from below'. For the kinds of difficulty – theoretical and methodological – both with this kind of social history and the resulting idealization of its subject, see Chapters 9 and 10 in this volume, and also Brass (2014b; and 2018a: Chapters 9 and 10).

Even with opponents it is possible to agree conceptually on the distinction between good historiography and, conversely, what of it is bad: this is very different from the position adopted by postmodern epistemology, which is underwritten by the claim that there *is* no history, nor can there ever be such a thing.[24] Instead, postmodernism attempts to convince that, because of aporia, all that remains is multiple fictions, none of which can be accepted as either true or false. History is reduced, in effect, to a form of novelization, a process uninformed by determination, all components of which – however minor – have to be regarded as equivalent in terms of epistemological value.[25] Negated thereby is not just the hierarchy of instances that structure Marxist theory about the history of development but also the notion of modernity itself, as an epoch linked to (and forming part of a) process that prefigured it. By challenging the conceptual efficacy of history, postmodernism amounts to a nihilistic declaration of war on reality, and with it any/every attempt to interpret it politically, let alone to build a programme on the basis of what history teaches.

III

Southern Myths

Set in North Carolina, Boston, and New York, over a period from 1865 to 1900, *The Leopard's Spots* by Thomas Dixon, Jr., is a fictional account of the Reconstruction era seen from the perspective of white Southerners on the losing side in the Civil War, a theme he continued in *The Clansman*, on which was based the film *Birth of a Nation*.[26] Each novel was published at the start of the twentieth century, and both proclaim the innateness of 'difference' based on race; the defeated South is depicted by Dixon as a place inhabited

---

24 Evidence of the kind of nothingness on offer can be found in what is an unabashed postmodern celebration by Munslow and Rosenstone (2004) of the attempt to 'rethink' history.

25 'The dizzying result', contend Munslow and Rosenstone (2004: 11, 14), 'is that as a literary (or filmic or hypertext?) form, History is now unknown territory … The specific object of experimental free fall is to force us to understand the past in new and different ways. To do this, the routine thinking and practice of "proper" epistemological History have to be made strange. And this can only be done by foregrounding the form of history as representation – literary, poetic, dramatic, filmic, and performative [hence] self-reflexivity is … the appreciation that history is a literary-creative act'.

26 The film *Birth of a Nation* (1915) was directed by D.W. Griffith. According to Dixon (1941: 1), *The Clansman* is part of 'a series of historical novels [about] the Race Conflict', and as such 'develops the true story of the "Ku Klux Klan Conspiracy" which overturned the Reconstruction régime'. The KKK is itself referred to as the 'Invisible Empire'.

by 'determined impassioned men [who] believed that this question was more important than any theory of tariff or finance [where the] most important development ... was the complete alienation of the white and black races as compared with the old familiar trust of domestic life'.[27] On this idealized view of pre-bellum South, where black and white are supposed to have lived in harmony, is then constructed a racist narrative, featuring betrayal by pro-abolitionist Northern whites of their pro-slavery Southern counterparts, which in turn licences and justifies resistance by the Ku Klux Klan, the actions of which are depicted in a similarly idealized form.

Depicted in positive terms by Dixon are three sorts of person: Southern whites, plebeian and aristocratic, all of whom served in the Confederate army; Southern blacks who are poor, but nevertheless continue subscribing to and defending the racist ideology of the antebellum South; and upper-class Northerners, some of whom served in the Union army, depicted as honourable. By contrast, those characters portrayed negatively also fall into three categories: emancipated black Southern ex-slaves, now freedmen, who though poor no longer accept their subordination, and challenge white supremacy exercised historically in the South; white Northerners who start out as poor, but enrich themselves by representing the political interests of still disenfranchised Southern blacks, alerting the latter as to their rights as citizens and contributing to struggles aimed at achieving them; and white Southerners labelled turncoats, who similarly enrich themselves as a result of having opportunistically transferred their allegiance from the defeated South to the victorious North for reasons of self-interest and personal gain.

Because he privileges race and a discourse about ethnic identity over political economy, Dixon – unlike Raper (see below) – attributes racism to an innate, unchanging and unchangeable concept of 'difference', a position linked by him in turn to sexuality. Although labour market competition as a contributing factor is not ignored, therefore, it is nevertheless subordinated epistemologically to a patriarchal narrative about protection by white males of the white female in what is presented by him as a traditional Southern

---

27 See Dixon (1903a: 159, 200). That race is the determining identity is evident from the kind of views expressed by characters who are white (Dixon, 1903a: 242, 331, 441): 'This racial instinct is the ordinance of our life. lose it and we have no future': 'Two great questions shadow the future of the American people, the conflict between Labor and Capital, and the conflict between the African and the Anglo-Saxon race. The greatest, most dangerous, and most hopeless of these is the latter': and 'I hate the dish water of modern world-citizenship. A shallow cosmopolitanism is the mask of death for the individual ... Race, and race pride are the ordinances of life'. While class struggle is acknowledged as a problem facing the nation, therefore, for Dixon it is overridden by the race issue.

code based historically on enduring religious and chivalric norms. Policing of female sexuality by Southern men is depicted as a crucial aspect in the preservation of racial 'purity', and thus to be deployed as a defence against proposed legislation forcing racial intermarriage.[28] Linking the issues of sexuality to the other main themes – the innateness of ethnic identity, Northern hypocrisy, and the commonality of white interests – Dixon shows that even a well-off Northern white who supports black equality and empowerment nevertheless disapproves of what is termed 'race-mixing' (= intermarriage) just as would his Southern counterpart, a combination illustrated in an episode when a well-educated (= 'acceptable') black man asks to marry his daughter.[29]

No Ear to Hear

As with gender, so with race. Since for Dixon the black population in the South is incapable of forming its own ideas, it is consequently perceived by him as gullible and taken in as to its interests by Northern members of the Freedman's Bureau.[30] What he objects to, therefore, is that blacks generally and ex-slaves in particular are being helped by Northerners to challenge the position and power of their erstwhile masters.[31] This is to be achieved by first enfranchising these hitherto subordinate and oppressed elements, and then utilizing the political majority thus created to 'confiscate the property of the [Southern] rebels' and transfer it to these now politically empowered categories.[32] Significantly, therefore, beneath Southern fears concerning the political

---

28   On these issues, see Dixon (1903a: 146, 333). Thus one character who is black, and described as 'the political boss of the new era' is given the following lines (Dixon, 1903a: 89): 'Our proud white aristocrats of the South are in a panic it seems. They fear the coming power of the Negro. They fear their Desdemona may be fascinated again by an Othello. Well, Othello's day has come at last'. Another character, who is also black, is lynched by the KKK because he asked a white female for a kiss (Dixon, 1903a: 147–50).

29   For this episode, see Dixon (1903a: 386ff.).

30   'If the devil himself had devised an instrument for creating race antagonism and strife', complains Dixon (1903a: 79), 'he could not have improved on this Bureau in its actual workings … [Its] agents were as a rule the riff-raff and trash of the North [who] were lifted from penury to affluence and power'.

31   A view expressed by Dixon (1903a: 46) thus: 'Your mission is to teach crack-brained theories of social and political equality to four millions of ignorant negroes … [y]our work is to separate and alienate the negroes from their former masters who can be their only real friends and guardians'.

32   See Dixon (1903a: 87–88).

enfranchisement of ex-slaves lies what is presented in the narrative (and by Dixon elsewhere) as being a threat to property relations.[33]

Hence the appeal by white Southerners to Northern equivalents is framed in terms of race (along the lines of 'ethnically we are alike'), emphasizing the sameness both of history (coming from a similar European tradition and heritage) and of kinship/religion/culture (analogous Puritan roots).[34] That white Northerners side with Southern blacks is depicted by Dixon as especially heinous, an egregious form of betrayal, its subject conceptualised negatively as a 'turncoat'. It is those in the latter category that attract the most vehement opprobrium, none more so than the character Simon Legree, reintroduced by Dixon into the narrative, and taken from its original appearance in *Uncle Tom's Cabin* by Harriet Beecher Stowe.[35] The characterization by the latter of Legree

---

33 This connection between racism, white solidarity, disenfranchisement of ex-slaves, and a danger to property relations was set out early on by Dixon (1889: 250) in the following manner: 'Why, then, do the young men, as well as the old men, stand as a unit in the determination that the negro shall not as yet control the local governments? Simply because they know he does not represent the wealth, virtue, and intelligence of the community, and because they know that negro supremacy in state or county means bankruptcy, ruin, disgrace, and corruption. I have no apologies to offer for the interference with the negro vote that has characterized certain sections of the South in times past'. In much the same vein he (Dixon, 1889: 251–53) continues by pointing out that ex-slaves 'took possession of the state government ... the land was taxed to nearly its full rental value, and the land was the only capital the people had [so the latter] took possession of their local governments [and] said it was not right that pauperism and vice and ignorance should rule wealth and virtue and intelligence ... This dark cloud of the possibility of local corruption and bankruptcy and ruin hangs like a pall over the South, and makes the white race stand together as a solid unit today. The enfranchisement of the negro race turned loose too much power'.

34 Labelling as a 'fanatic' a well-meaning and wealthy female from the North intent on promoting the emancipation and well-being of ex-slaves, a Southerner asks her (Dixon, 1903a: 47–48) the following question: 'Why is it that you good people of the North are spending your millions here now to help only the negroes, who feel least of all the sufferings of this war? The poor white people of the South are your own flesh and blood. These Scotch covenanters are of the same Puritan stock, these German, Huguenot and English people are all your kinsmen, who stood at the stake with your fathers in the old world. They are, many of them, homeless, without clothes, sick and hungry ... But one in ten of them ever owned a slave. They had to fight this war because your armies invaded their soil. But for their sorrows, sufferings and burdens you have no ear to hear and no heart to pity. This is a strange thing to me'. To this she replies that the South can look after its own, having been punished by God for owning slaves and fighting against the nation.

35 On the character of Legree, see Stowe (c. 1860: 68ff., 199ff.). As portrayed by Dixon (1903a: 84–85), Legree reappears, having migrated north: 'The rumours of his death proved a mistake. He had quit drink, and set his mind on greater vices ... When the war closed, [he] became a violent Union man, and swore that he had been hounded and

as the epitome of a 'bad Southerner', a planter who oppresses and kills his slaves, is turned by Dixon on its head: Legree reappears in the narrative as an instance of a 'bad Northerner', one who from a Southern viewpoint exemplifies an unprincipled opportunist and thus the worst kind of turncoat.

To this end, the murderous villainy carried out by Legree is transferred from a form of notoriety associated only with the South and affixed instead to the North, the latter being the area to which the former planter flees; to this already negative image Dixon adds peculation as the result of holding political office, enabling the character to build a personal fortune.[36] Having embezzled public funds, Legree is described as all-powerful, owning factories in New Jersey where trade unions are forbidden.[37] If Legree deserved opprobrium for his treatment of slaves in the South, implies Dixon, then how much more blameworthy is the way he treats his free labour in the Northern industries now owned by him. This amounts to a thinly disguised critique of Northern capitalism as systemically negative when compared to what is presented as the systemically benign ante-bellum plantation with its slave workforce. Needless to say, the latter claim is a staple of Southern ideology: pro-slavery discourse combined with opposition to capitalism, or the anti-capitalism of the political right.[38]

Against the Northern argument that 'industrial training' (= work discipline) will decrease the dangers feared by Southern whites regarding the effects of black emancipation, Southerners maintain that 'industrial training gives power': hence the Northern view is rejected by white Southerners who warn that ex-slaves will henceforth be able to exercise workplace solidarity as members of the working class. Hinted at by Dixon is the underlying concern on the part of better-off whites that blacks may even as a consequence unite with

---

persecuted without mercy by the Secessionist rebels ... He was a little older looking than when he killed Uncle Tom on his farm some ten years before, but otherwise unchanged'.

36   'Simon Legree is more than a mere man who stole five millions of dollars, alienated the races, and covered the South with the desolation of anarchy', is the opinion of him as expressed by a white Southerner (Dixon, 1903a: 195), since '[h]e represented everything that the soul of the South loathes, and that the Republican party has tried to ram down our throats, Negro supremacy in politics and Negro equality in society'. In this way the South is distanced from the negative image that Legree represents, by portraying him as a turncoat who becomes a Northern carpetbagger.

37   Asking 'does this fiend exite the wrath of the righteous?', Dixon (1903a: 399) answers: 'Far from it. His very name is whispered in admiring awe by millions ... His name is magic', thereby blaming Northerners generally and Northern capitalism for approving both the way Legree conducts his business and also the kind of person he is.

38   See Dixon (1903a: 398–99, 401). On the anti-capitalism of the political right, and why it is important to distinguish this from its leftist variant, see Brass (2022b: Chapter 2).

white labour, posing a threat to hegemony based on race, and ultimately to property relations in the South. However, upholders of Southern 'difference' take refuge in the divisive effect of labour market competition as capitalist development proceeds, countering any possibility of class solidarity overriding that based on race.[39] In words given by Dixon to a Southern protagonist, 'If the Negro ever becomes a serious competitor of the white labourer in the industries of the South' lynching will follow.[40]

Differences, Sameness

Accordingly, at the end of this development trajectory, argues Dixon via his Southern voices, lies a political transformation that threatens the private property of Southerner and Northerner alike: namely, a transition to socialism.[41] As such, he infers, it is yet one more reason for white unity across the Civil war divide. A portent of this outcome, feared by planter and industrialist, is the Freedman Bureau programme of agrarian reform, whereby emancipated slaves are to receive land expropriated by the state from their masters.[42] The

---

39 That potential/actual labour market competition involving different ethnic groups, and its contribution to the racism of white plebeians, was an issue that surfaced regularly throughout pro-slavery discourse in the antebellum South, is evident from the contributions to Elliott (1860). As the analysis of Raper (see below) indicates, this became more acute as capitalism spread, particularly so during the Great Depression.

40 Asked 'will not your industrial training of the Negro gradually minimize any danger to your [Southern] society', the Southerner replies (Dixon, 1903a: 335): 'No, it will gradually increase it. Industrial training gives power. If the Negro ever becomes a serious competitor of the white labourer in the industries of the South, the white man will kill him, just as your labour unions do in the North now where the conditions of life are hard, and men fight with tooth and nail for bread'. For more on this outcome of labour market competition between black and white, see Dixon (1903a: 381, 396–97, 402, 459).

41 On the threat of socialism linked to the combination against which Southerners fought – slave emancipation, black empowerment, and capitalist development – see Dixon (1903a: 397–98). This same argument, that capitalism leads to socialism, is made in other novels he published (Dixon, 1903b; 1909; 1911), one of which – *Comrades* – was turned into a silent film, *Bolshevism on Trial* (1911), directed by Harry Knoles.

42 Giving rise to a situation of which Dixon (1941: 238, 246) clearly disapproves: "'A great day for our brother in black. Two years of army rations from the Freedman's Bureau, with old army clothes thrown in, and now the ballot – the priceless glory of American citizenship. But better still the very land is to be taken from these proud aristocrats and given to the poor down-trodden black man. Forty acres and a mule – think of it! ... your old masters to work your land and pay his rent in corn, while you sit back in the shade and see him sweat'".

latter has clear implications for one main Southern protagonist, a businessman combining agricultural production with the manufacture of cotton, who strongly opposes land redistribution and the franchise leading to this, in the form of black voting rights.[43] His solution is twofold: emphasizing how much better off blacks were as slaves on the antebellum plantation, or the aristocratic pastoral version of agrarian myth discourse; and calling for the expulsion of ex-slaves, sending them to Africa.[44]

That Dixon subscribes to the aristocratic pastoral version of agrarian myth discourse is unsurprising, for two reasons in particular.[45] First, he espoused an ideology about the rural nature of the South, at the centre of which was the image of a benign planter class providing for the needs of its unfree workforce and simultaneously caring for the wellbeing of plebeian white farmers. Hence the concern felt by a landowning Southerner about the threat represented by the agrarian reform policy of the Freedman Bureau plus wealthy Northern carpetbaggers to peasant smallholdings cultivated by whites.[46] And second, as outlined in his autobiography, he himself owned Elmington Manor, a two-hundred year old 'stately colonial home' on an estate consisting of 500 acres,

---

43   'The man who tells you that your old master's land will be divided among you, is a criminal, or a fool, or both', the white Southern businessman tells ex-slaves (Dixon, 1903a: 66–67), adding: 'The man who tells you that you are going to be given the ballot indiscriminately with which you can rule your old masters is a criminal or a fool, or both'.

44   That a discourse combining the agrarian myth, racism, and far-right politics was at this conjuncture not confined to the American South or Dixon is evident from the case in England of Cecil Chesterton. Not only did his 1919 account of the history of the United States (Chesterton, 1940: Chapters 8-10) echo the arguments made by Dixon, but he influenced both the pro-peasant Distributist views of his brother G.K. Chesterton, and the far-right ones of his first cousin, A.K. Chesterton.

45   Central to the agrarian myth – on which see Brass (2000) – is a pro-rural/anti-urban discourse that is also given expression in Dixon (1905: 138): 'This acme of living cannot be attained in the city ... I believe that man's full growth will be best reached by spending one-third of his time in town and two-thirds in touch with Nature .... The gravest charge against the modem city is not merely that its continuous unrest starves the soul, brutalizes the senses, destroys repose, develops insolent and savage impulses – but worse than all this, it murders the will and destroys personality, thus sapping the fountain of life. The collective instincts of the herd-groups in which we move strangle at last the individual, and man becomes but a grain of dust blown hither and thither by the breath of a crowd'.

46   This issue is presented by Dixon (1941: 189) in the following manner, uttered by a Southern woman about to be deprived of her property: "'Just as I was about to give up – the first time in my life – here came those rich Yankees and with enough to rent to pay the interest on the mortgages ... last week I cried when they told me I must lose the farm [that] was my dowry with the dozen slaves Papa gave us on our wedding-day. The negroes did as they pleased, yet we [the white owners] managed to live and were very happy'".

of which 350 were under cultivation, situated near Chesapeake Bay.⁴⁷ Built by slaves, the views Dixon held about black labour are to some degree based on his own experiences as an employer.⁴⁸

Expelling blacks from America, by sending them to Africa, is a policy that has a familiar contemporary ring to it. Unsurprisingly, therefore, echoes of what is now seen as the rightist discourse termed 'great replacement' abound in the narrative of Dixon, where it features as an ideology invoked not just by Southern whites to justify their own racist views about blacks, but also by the latter and their Northern supporters, so as to lend credibility to the espousal of this discourse by white Southerners.⁴⁹ Thus a white smallholder is said to hate blacks because they represented for him a rival force on the plantation that made life difficult for people like him, 'crowdin' us to death'.⁵⁰ Later on, another character states that '[w]e will drive the white man out of this country. That is the purpose of our friends at Washington', adding: 'If they don't like their job [being servants to blacks] they can move to a more congenial climate'.⁵¹ Much the same kind of view surfaces regularly throughout the account by Dixon of what are presented as Northern threats leading to Southern fears.⁵²

---

47 This rural location is depicted by Dixon (1905: 7) thus: 'And then the longing for the country life in which we had both been reared came over us with resistless power. The smell of green fields and wild flowers, the breath of the open sea, the music of beautiful waters, the quiet of woodland roads, the kindly eyes of animals we had known, the memory of sun and moon and star long lost in the glare of electric lights, began to call'.

48 'I had never believed the wild stories about the modern negro farm labourer in the South till I tried it', contends Dixon (1905: 48), explaining that: 'In three years I've hired over one hundred negro farm hands and discharged all save three of them, who are first-class men. I tried patiently to teach one I kept six months to do a few simple necessary things with modem farm machinery. At the end of six months he broke three mowers in one day on a beautiful level piece of clover'.

49 On the 'great replacement' approach, see Brass (2022a).

50 '"I always hated [blacks] since I was knee high"', this plebeian character says (Dixon, 1903a: 28), adding: '"My daddy and my mammy hated'em before me. Somehow we always felt like they was crowdin' us to death on them big plantations, and the little ones too"'. The same notion concerning status reversal and ethnic replacement surfaces elsewhere, when a black character states (Dixon, 1941: 205): '"Give the white trash in this town to understand that they are not even citizens of the nation. As a sovereign voter, you, once their slave, are not only their equal – you are their master"'.

51 See Dixon (1903a: 91, 95), who puts the same great replacement discourse in the mouth of a preacher addressing a church congregation, telling the latter that '[t]he attempt is to be deliberately made to blot out Anglo-Saxon [= white] society and substitute African barbarism'.

52 See, for example, Dixon (1903a: 242, 335), who makes this fear/threat connection evident in the following manner (Dixon, 1903a: 438–39): 'We grant the Negro the right to life, liberty and the pursuit of happiness if he can be happy without exercising kingship over the

IV

Common Heirs to Its Impositions

Turning from the image of race depicted in the pro-Southern narrative of Dixon in his fiction to the sociological analysis of the way the same identity featured in the same context before and during the Great Depression of the late 1920s, a rather different picture emerges. In the period after the Civil War, many plantations in the South were subdivided into tenancies and smallholdings.[53] Over the 1885–1900 period European immigrants arriving in the rural South facilitated landlord restructuring along ethnic lines, by replacing black tenants with white equivalents and introducing more of the latter into what had been previously a black area, 'complicating the race situation'.[54] Tenants who are white not only have larger holdings, but are consequently in a position to buy them at some future date, enabling a transition to small landed proprietors, and thus owners of the means of production. The subsequent migration of landless blacks into such areas meant that white smallholders were in

---

Anglo-Saxon race, or dragging us down to his level. But if he cannot find happiness except in lording it over a superior race, let him look for another world in which to rule. There is not room for both of us on this continent'.

53 Of the two main sources used here for an account of Southern rural society over a period from the late 1920s to the early 1930s, one refers to the Black Belt while the other about what occurred more generally throughout the South. Hence methodologically the former (Raper, 1936) draws upon data from interviews conducted with 300+ families in Greene and Macon counties, Georgia, in 1927 and again in 1934 (Raper, 1936: 33), while the latter source (Raper, 1970/1933) analyses the incidence and causes of lynching in the South during 1930. An academic sociologist who was also a member of the Southern Commission on Interracial Cooperation, Arthur Franklin Raper (1899–1979) undertook extensive research into the causes and effects of racism and rural poverty in the Depression-era South, examining issues such as economic development in the countryside, employment conditions, sharecropping, wage labour, migration patterns, and tenure structure, together with how and why they were all interconnected.

54 See Raper (1970: 100, 130), who comments: 'An interesting landlord-tenant experiment is now under way, where a leading white planter is using white tenants. Last year he had on his plantation 125 white families and one Negro family. Ten years ago he was using only Negroes. This year, he will use fewer than 100 families, because of the greater acreage each family will cultivate ... This experiment is of significance for several reasons: it has brought additional tenant whites into this Negro tenant area, immediately complicating the race situation; it is based on a type of farming which makes possible a cash surplus for tenants; and it is making small farm ownership available to white people, thus affording the possibility for the development of a middle-class white element'.

competition with incoming ex-slaves and ex-sharecroppers, not just for land but also for work in the off-season, when impoverished rural elements – white and black alike – relied heavily on income from selling their own labour-power in the same market.⁵⁵

By 1930, over ninety percent of the black population who owned no land in rural areas were tenants, compared with only half of those who were white and landless.⁵⁶ Given the post-bellum insertion into the Southern rural labour market of ex-slaves and whites, all of whom were both landless and thus possessing as their sole commodity for sale their labour-power, it is unsurprising that not only was racism fuelled by more acute competition between black and white plebeians for land and employment, but also that such workforce division was itself promoted by industrial and agrarian capitalists, as employers drew on the surplus labour released from an agriculture undergoing economic crisis.⁵⁷ Although the main beneficiaries of a fragmenting plantation system were poor whites who became sharecroppers or smallholders, nevertheless like their landless black counterparts their income relied substantially on wages received from the sale of their own labour-power.⁵⁸

This historical context in turn foregrounded the racial dimension of competition between black and white landless workers for tenancies that became more acute during the 1930s, not least because those in the former category

---

55 As plantation agriculture in one part of the South contracted, its erstwhile workforce moved to other rural locations, the result being that in such contexts (Raper, 1970: 140) 'the masses of whites have always been in competition with Negroes and antagonistic toward them. The recent low prices of tobacco, melons, and peanuts have destroyed the hope that these new crops would restore the short-lived prosperity of earlier years'.
56 On this background see the account by Raper (1936: 143ff.), where he observes that: 'In the Black Belt plantation area of the lower South there are not only more landless farmers than elsewhere in these states, but a greater percentage of the landless farmers, white as well as Negro, are croppers and wage hands'. Data for 1930 provided by Woofter (1936) arrive at a lower figure, indicating that 58% of Southern blacks were tenants, 29% were labourers, and 13% were peasant proprietors.
57 See Raper (1936: 5–6), who comments that: 'To no small degree the Black Belt is the seed-bed of South's people and her culture. Human relations in Atlanta, Birmingham, Montgomery, Memphis, New Orleans, and Dallas are determined largely by the attitudes of the people of the Black Belt plantations from which many of their inhabitants, white and Negro, came ... No real relief can come to the region so long as the planter, who wants dependent workers, can confound the situation by setting the white worker over against the black worker, and so long as the industrialist, who wants cheap labour, can achieve his end by pitting urban labour against rural labour. There are literally millions of farm labourers in the Black Belt who are eagerly awaiting an opportunity to work for wages even smaller than are now being paid textile and steel workers in southern cities'.
58 Raper (1936: 186).

were preferred by Southern landlords since they were deemed to be 'more easily supervised' and 'do not advance into ownership as readily as white tenants'.[59] Like privately owned smallholdings, tenancies were also a source of labour market competition, insofar as there were present also on rented land resident sharecropper kinfolk 'who do work for wages now and then throughout the year'.[60] Accordingly, it is impossible to decouple these two seemingly unconnected forms of struggle: that for access to land – either as a peasant proprietor or sharecropping tenant – and for waged labour, contributing thereby to increased competition in the search for jobs. As well as conflict with the owners of the means of production, therefore, the landless of any ethnicity also faced one another as rivals in the labour market, a struggle that gave an impetus to the notorious and racist practice of lynching.

The Battle for Bread

For those who either lose their livelihood and become unemployed, or whose employment or tenancies were under threat, as a result of the late 1920s capitalist economic crisis in the South, lynching was a method of expressing hostility/antagonism towards actual/potential rivals in the labour market. Cut-backs in industrial production during the Depression-era led to a reverse migration process, as many of those who had gone north in search of work then returned to their original rural locations in search once again of work prospects there, generating yet more intense competition for land and rural jobs in what had earlier been sending areas of the South.[61] Thus the Great Depression intensified labour market competition, as poor whites now sought employment opportunities previously held by blacks, an economic rivalry the latter referred

---

59 See Raper (1936: 148, 149), who underlines the element of ethnic competition for access to land: 'The presence of landless Negroes in the Black Belt does not improve the condition of the average white man in relation to the soil. The percentage of white tenancy is almost as high in the Black Belt as in other parts of the state, and the proportion of white tenants who are in the lowest tenure class croppers – is higher in the Black Belt than elsewhere. Since the plantation system forces white tenants and Negro tenants into competition, there is but little possibility of either group rising'.

60 Raper (1936: 153). In the Black Belt during the early 1930s (Raper, 1936: 157), '[t]he relation of landlord and tenant ... retains something of the quality of the master-slave régime. The increase of white tenants in recent decades has done little to modify the traditional patterns which grew up along the colour line in the years following the civil war. The white and Negro tenants, competing within the system for farms, are now the common heirs to its impositions'.

61 On this point see Woofter (1936).

to as 'a long standing "battle for bread"' between plebeians of different ethnicities.[62] Blacks were told to leave areas within 24 hours – that is, quit their jobs and move elsewhere – while employers were ordered to replace black workers with whites, an instruction that was followed.[63]

Although landlords were not slow in promoting ethnic division, turning it to their economic advantage, at times they themselves were the target of hostility from white plebeians. On occasion, therefore, the antagonism of poor whites was aimed at rich counterparts, on the grounds that the latter – as landlords – either demonstrated a preference for employing blacks as tenants or workers, or enabled them to escape the retribution of lynch mobs.[64] In keeping with this is the difference in attitudes consequent on the erosion of white middle class: one small component accumulated property and rose up the social hierarchy, while the larger part moved down the class structure, becoming tenants and being no different from poor whites. Those who moved up tended to paternalistic views on race, exhibiting less antagonism towards blacks, while those who moved down into tenancy did not, and supported lynching.[65] These

---

62   'The uncertainty of employment and the slump of farming had further embittered the propertyless whites who openly begrudged the Negroes any evidence of accumulated savings or of regular employment', comments Raper (1970: 340), who illustrates this contention as follows: '[T]he bellboys in the Grayson Hotel had been displaced by white boys about 1928, while within the six months previous to the disorders of May, 1930, white people had applied for the places of many Negro janitors [who] spoke of the riot as a phase of a long standing "battle for bread" between the poorer whites and the Negroes'.

63   Raper (1970: 341). For similar instances in the South at this conjuncture, see Raper (1970: 313, 317ff.), who argues that 'in many communities the organized efforts of whites to displace Negro laborers with unemployed whites may be expected to aggravate racial animosity to the level of open conflict and violence'. (Raper, 1970: 31).

64   On this see Raper (1970: 202, 259–60, 262, 276, 281–82, 313), who comments: 'The poorest whites and Negroes, working as farm wage hands, propertyless tenants, and at sawmills, are in daily competition for the necessities of physical existence. The few employers of labor, however, get the most labor they can for their money, and so the Negroes eke out a living along with the poorer whites' (Raper, 1970: 171).

65   In what is a view about upwards/downwards mobility that is similar to that of Lenin regarding peasant differentiation in pre-1917 Russia, the fragmentation of the white bourgeoisie in the 1930s American South is described by Raper (1970: 193) in the following terms: 'Although this increase of Negroes and outside whites has modified the whole social and economic structure, there still remains a vestige of the old class distinctions as between the English planter class, the middle class whites of Scotch descent, and the poorer whites. These old class lines have been modified further by the virtual disintegration of the white middle class ... A part of it has accumulated property and become identified with the old landed aristocracy; a larger proportion has slumped into tenancy and become identified with the poorer white element. The Scotch descendants who rose into the wealthy class have been friendly to Negroes ... On the other hand, the Scotch descendants who passed into the tenant class tend to justify or condone [lynching]'.

were precisely the elements most in competition with blacks entering the labour market.

Hard to avoid, therefore, is the similarity in terms of class position of those composing both the white lynch mob and also its black victims. Throughout the South in 1930, therefore, lynch mobs were generally comprised of unemployed and propertyless young white males, whilst in rural communities participants in such episodes consisted of farm tenants, landless farmers, and wage hands.[66] Among mob leaders were 'an itinerant drunken roustabout', 'home owners and traders in livestock', and 'a plantation wage hand' described by his landlord as 'a hard-working and honest man of the landless, shifting type', while the lynch mob itself 'was made up from the lowest elements of the white population'.[67] Lynchers included those who were 'propertyless', 'unemployed', and 'very poor'; none of them were said to 'own any taxable property'.[68] Lynch victims are described variously as coming from black plebeian strata: 'a wage hand ... a hard-working field hand'; a sharecropper who 'hardly broke even'; a black worker with a road construction gang; 'an itinerant labourer'; a worker that 'went to his employer's house asking for wages'; and an 'itinerant, illiterate labourer [who] had been living in [the area] but a few months [described by his former employer as] hard-working and the best help about the farm he had ever had'.[69] That another victim 'had been in the county only a few weeks' underlines the extent to which the targets of lynching were incomers/newcomers either looking for work or having just found employment.[70]

The Pinch of Hunger

Lynching was but one aspect – the most notorious part – of an ensemble of coercive practices operating in Southern agriculture.[71] Designed to retain

---

66 See Raper (1970: 11, 186). Those indicted for lynching included textile workers, farmers, a blacksmith, an ice plant employee, and a filling station operator (Raper, 1970: 274).

67 Raper (1970: 243–44).

68 Raper (1970: 331).

69 For these cases, see Raper (1970: 144–45, 286ff., 289, 306, 319, 329–330). The thesis he advanced linking an enhanced incidence in lynchings to increased economic competition in the labour market has been upheld by subsequent research. A rise in the number of blacks lynched by whites is confirmed as having occurred during peak labour demand for cotton and tobacco crops (Ayers, 1984; Beck and Tolnay, 1992; Tolnay and Beck, 1995), a finding that is itself supported by even more recent analysis (Christian, 2017).

70 Raper (1970: 334).

71 Included among these coercive practices was disenfranchisement; to prevent blacks from exercising power, and being able politically to challenge their discriminatory and

employer access to the labour-power of the tenant family, by preventing its components from entering the market for this commodity, the system of tied housing in the rural South at this conjuncture ensured control of the workforce on the plantation, reproducing thereby the continuation – albeit in a different form – of the unfree production relation of the slave era.[72] As in the case of Britain during the nineteenth century, therefore, unfreedom in the rural South was linked to housing. In spite of being able still to sell their own labour-power to other employers in the vicinity, even hired labourers were engaged on a beck-and-call basis, enabling them to work for others only when not required by the farmer on whose land they resided.[73]

Hence the ever-present threat of eviction from where tenants, hired workers, and their families lived in effect secured their labour-power for the planter concerned. The object of tied housing in reproducing workforce unfreedom meant that only white and black tenants and hired labourers alike who, because they owned their houses, were as a consequence able to sell their personal labour-power to whomsoever they wished. Females in the tenant household were required to work on the tenancy leased from the landlord, whereas the sons worked as casual labour on a daily or weekly basis elsewhere whenever such employment is available. Those in the former category are unfree, while those in the latter are free wage labour.[74] Moreover, since the female labour-power of the tenant family was unfree, its availability for sale in the labour market was anyway confined to periods when there was no demand for this commodity.[75] Because its deployment was already spoken for, to be applied

---

oppressive situation, whites prevented them from voting in elections, a procedure euphemistically known as 'disturbing the count' (Raper, 1936: 165ff.).

72 See Raper (1936: 154), who observes that '[i]n the rural Black Belt, except for casual laborers, [employment] is a yearly proposition, for even the wage hand, though paid only for the days the landlord wants him, must be available at all times and it is on this condition that he is furnished a farm cabin in which to live'.

73 On the role of tied housing in the enforcement of unfree production relations in British agriculture from the nineteenth century to the present, see Brass (2004).

74 The interconnectedness of all these issues is emphasized (Raper, 1936: 155) thus: 'The Negroes and whites living in the country who own but a house and a lot can apply their labor where they choose only because they own their living quarters and do not own any land. The share renters provide a second source of casual day laborers; for while the head of the house and the women work at the crop, the sons of the family not infrequently work by the day or week wherever they can find employment'.

75 As '[t]he fixed-renter [sharecropping tenant] is dependent upon the application of his labor to the tract of land he rents', notes Raper (1936: 154–55), '[t]he only time a cropper's family has labor for the open market is when there is but little demand for it'.

only on the tenancy cultivated by the household head, outside earnings and consequently family income as a whole remained low.

The outcome of this combination – landlessness, unfree production relations, few outside earnings, and low income – meant that competition both for tenancies and for labour market access became a paramount consideration for buyers and sellers of labour-power alike: for workers in order to meet their subsistence requirements, and for landlords seeking to recruit yet constrain/control them. For their part, workers regarded their poverty as an effect of this combination which, being anyway landless also reduced or prevented commodification of their only remaining commodity, thereby reproducing their economically dependent status *vis-à-vis* the planter class.[76] By contrast, landlord discourse not only identified hunger as the only true guarantor of labour commodification ('the pinch of hunger'), but also attributed the poverty of the Southern tenant and worker to improvidence and indolence ('they only work when they have to'). The wife of a tenant or labourer will drive him to work, so this same discourse avers, only when there is no food in the house, so she – like him – is to blame for their poverty, the result of fecklessness on the part of each.[77]

Debt peonage was a related form of coercion operated and enforced by landlords throughout the South.[78] Hence an indebted worker who absconds without repaying what is owed risks landlord retribution being visited on his family and other kinsfolk.[79] This sort of compulsion is linked to another form

---

76 On the low level of cash incomes of those working for wages, see Raper (1936: 39–40, Tables v and vi).

77 Shortcomings of the rural worker as perceived by the Southern landowner adhere to all the familiar racist/gender stereotypes, and take the following form (Raper, 1936: 158–59): 'The typical wage hand or tenant, asserts the typical landlord, will work when he has to – when he is hungry or about to get hungry; his improvidence is so thorough that the little money in hand causes him to work irregularly or not at all. His wife, too, it is said, will help him get off to work if there is little or no food in the house, not so if there are provisions on hand for more than a few days; the pinch of hunger, as nothing else, will get the shiftless, child-like tenant into the fields and keep him there'.

78 On debt peonage in the rural South, see not only Raper (1970: 83, 127) but also Kennedy (1946), Daniel (1972), Lichtenstein (1996), and Angelo (1997).

79 'This practice is rendered effective', notes Raper (1936: 172), 'by a kind of gentleman's agreement among the planters that they will accept a wage hand or tenant from another plantation only when the change meets with the approval of his landlord, in which case the new landlord assumes his indebtedness to the old, applying it against the tenant's future crops ... Thus robbed of his normal right to move, the propertyless farm worker can escape his lot only by fleeing the community, which involves considerable risk. Threats of flogging, murder, and lynching follow and sometimes overtake the "debtor" who "slips off like a thief." If he gets safely out of reach, the violence originally meant for him may

of coercion, whereby a labourer owing debt to one landlord is permitted to find employment with an alternative proprietor only on condition that the debt accompanies the worker concerned, a transaction (= 'changing masters') amounting to inter-employer transfer of debt which merely shifts the coercion/unfreedom relation from one creditor-employer to another.[80] Landlords maintained that such control, and coercive practices more generally, were necessary because '[t]he very life of the plantation system is threatened when tenants accumulate property, exhibit independence'.[81] Consequently, in the struggle between landlords and their workers, negative views held by the former about the latter 'are the rationalizations and defence mechanisms which the controllers of the plantation system have fabricated into a philosophy which justifies and maintains the politically sterile "Solid South" and its outmoded agricultural structure based upon the human relations of a disintegrating feudalism'.[82]

Conclusion

About racism, why/when it surfaces, why it endures and how it is reproduced, therefore, the position Marx holds is that unless the underlying economic conditions are addressed, and then eliminated, the ideology itself will endure. This interpretation is consistent not just with the approach of Engels and later Marxists (Lenin, Trotsky, Deutscher) but also with the argument made here regarding the way labour market competition generated by capitalism – an integral systemic aspect, and one on which accumulation depends in order to keep its costs down and maintain profitability – intensifies and sustains racism. This is perhaps the very systemic contradiction – the antimony between the needs on the one hand of accumulation and on the other of society at large – that Marx envisaged as being the one that capitalist development can in the end neither avoid nor evade.

The connection between race and labour market competition surfaces in contrasting ways the American South is depicted in the fiction of Dixon and the sociology of Raper. The fiction of Dixon and the film *Birth of a Nation*

---

descend upon his relatives and friends, for little mercy is shown to those who dare challenge the plantation controls'.
80 On the presence and reproduction by capital of this kind of unfree labour relation ('changing masters') in other contexts, see Brass (1999).
81 Raper (1936: 171).
82 See Raper (1936: 170–71).

depicts race in the post-bellum Southern states during the Reconstruction era as a question mainly of sexuality, as a perceived threat by black males to white females, whereas by contrast the sociological research conducted by Raper at that conjuncture presents the same issue in different terms, as one of labour market competition. Each addressed the same issue: how and why ideas about race were generated and reproduced, but with distinct and politically opposed ideological frameworks.

Hence the racist narrative of Dixon is based on ethnicity as an identity that is historically innate and unchanging, a discourse he combines not just with theory about great replacement of whites by blacks but also with an image of lynching as a white plebeian variant of the agrarian myth, in which the embattled pastoral becomes the Darwinian variant. This discourse conceptualises both females and ex-slaves in patriarchal terms: blacks are infantilized, which requires in turn that agency undertaken by them is an effect of their manipulation and deception by white Northerners. Linked to this is the claim made by Southerners about Northern hypocrisy, in that Northerners condemn the South for its slavery, while praising the endeavours of capitalism in the North, despite the claim by the Southerner that the workforce fares much worse economically in the free market of the North than in the South, where labour-power is unfree.

This is the anti-capitalism of the right, an ideological trope of Southern discourse, which claims the slave was fine under the plantation system during the antebellum period, when all his own wants and those of his family were met, whereas in the free market of Northern capitalism no livelihood is guaranteed, and the worker has to fend for himself as best he can. According to this same discourse, black ex-slaves who come North merely compound the difficulty faced by white labour, since they enable producers to exert further downward pressure on existing wages and conditions, by adding to the quantity of labour-power entering the employment market, to the economic benefit of capital.

A central issue to emerge from sociological research conducted by Raper in the 1930s American South, however, is not just that at the root of racism is labour market competition between impoverished whites and landless blacks, but also the contributory role to this divide on the part of wealthy landowning whites. Underlined thereby is how on occasion what might have united ethnically distinct components of the same class (blacks and whites separated from their means of labour) is in the course of struggle replaced by antagonism derived from these different racial identities (the same components appearing simply as black and white rivals in search of the same land and work). Whether it was landless whites or landless blacks that entered the labour market, the result was the same: either to defend their own precarious access to land as tenants or to employment as seasonal labour, whites used racist ideology and lynching in order as they saw it to keep blacks from entering the labour and land market or – if they were already in there – expelling them from tenancies and jobs.

Because he subscribed to agrarian myth ideology, and came from a landowning family, Dixon also interpreted white Southern identity in terms of agricultural interests. As a landowner, therefore, the main threat was to rural property, and would come both from ex-slaves who were tenants or landless workers, and from socialism. As important was the threat to the peasant family farm, on which landowners relied for peak season labour supplied by a workforce composed of kinsfolk, including mothers, daughters, and sisters, on whose labour-power a smallholding depended, and the loss of which put its – and therefore his – economic reproduction at risk. However, the threat *to* workers, as distinct from one *by* them, was secondary. For the sociologist Raper, by contrast, labour-power itself was the main kind of property at issue between blacks and whites. Although like Dixon, racial antagonism was viewed as an effect of conflict over labour-power in particular rural contexts, sociological research presented the issue much rather as one involving the availability, sale and purchase of this commodity during a period of acute economic crisis.

Finally, where race is concerned, a theoretical/ideological contrast separates the political economy approach both of Marxism and of sociological research conducted in the postbellum South from the culturalist one of Dixon and postmodernism. Omitting to historicize the way ethnicity/nationality is conceptualized, changes, and is reproduced over time (let alone by who and why), postmodernism essentialises such non-class identities, regarding them as innate and empowering. Among the less than benign results are two in particular. First, the privileging of those components of traditional culture to which Marxism has always been opposed; and second, a position taken by postmodernism on the industrial reserve that is no different from that held currently by capitalists, one the object of which is ironically to disempower the very grassroots subjects whose interests postmodern theory claims to represent. As will be seen in the following chapters, the implications of this epistemological and political divide between Marxism and postmodernism for the study of development are profound.

CHAPTER 2

The Industrial Reserve and Development: a Vanishing Army?

> And the migrants streamed in on the highways ... They had no argument, no system, nothing but their numbers and their needs. When there was work for a man, ten men fought for it – fought with a low wage. If that fella'll work for thirty cents, I'll work for twenty-five. If he'll take twenty-five, I'll do it for twenty. No, me, I'm hungry. I'll work for fifteen.
>
> A description by JOHN STEINBECK of the way labour market competition operated in America during the Great Depression.[1]

∴

> In order to maintain car production [at the Renault factory] in France, the employees had to redouble their efforts, show that their expertise was more valuable than low-cost manpower abroad. The consequence of this fierce competition was an ever-greater exploitation of all workers: those who wanted to keep their jobs, and those who hoped to get one. Both sides lost.
>
> A description by DAVID FOENKINOS of the way labour market competition operated in France some eight decades later.[2]

∴

Introduction: Redefining the Industrial Reserve

Socialists have long argued that the industrial reserve army is one of the most powerful weapons in the economic armoury at the disposal of capital in its

---

[1] Steinbeck (1939: 343).
[2] Foenkinos (2020: 217–218). Although both epigraphs are drawn from writers of fiction, each nevertheless manages to encapsulate the same impact of labour market competition – eighty years apart – on workers employed by capital in different contexts.

struggle with the working class. As capitalism goes global, so too does this method deployed by producers to keep down labour costs, discourage unionisation, and the emergence or consolidation of class solidarity. Accordingly, an expanding industrial reserve army, generated as a result of globalization, enables capital to adopt a divide-and-rule policy, which involves turning workers of different non-class identities (ethnicity, nationality, gender) against one another in their attempts to secure employment. It has become a common tactic utilized by employers in a *laissez faire* economic context where capital has unfettered access to what is now a global industrial reserve army, giving additional momentum to the 'race to the bottom' in pay and conditions. This is especially true of contexts where migrants are many and jobs are few, a situation when competition in the labour market develops in a particularly intense manner.

Given this negative history in relation to the political economy of capitalist development, why then has the industrial reserve either vanished from consideration by social science discourse, or else – much the same thing – been redefined in positive terms? How is it that a concept that Marxism regards as crucial to an understanding of the way in which class struggle fails or succeeds, one that now is more relevant than ever as capitalism has become a systemically global phenomenon, no longer sparks the kind of political interest it once enjoyed? In an attempt to address these questions, therefore, the object here is to examine and account for the different outcomes of contrasting approaches to the industrial reserve, both within and outside Marxist theory.

The ambivalence regarding surplus labour, combining a recognition of its negative aspects, an acceptance of its importance for the accumulation process with an inability to dispense with this economic role, underlines the dilemmas facing liberal defenders of capitalism, extending from Beveridge a century ago to Fukuyama now.[3] For their part, those who address what Marx said about the industrial reserve, fall into one of two categories. First, upholders of the view that what Marx said originally about its negative political and economic impact on working class solidarity (= divide and rule tactic) still applies. And second, an approach that maintains what Marx really meant was the opposite, in that the surplus labour composed of migrants is not an obstacle to working class empowerment but much rather is today found to be in the vanguard of the struggle against capital: for this reason, therefore, the industrial reserve should be viewed positively.

---

3 For the difficulties facing the current attempt by Fukuyama to address – never mind resolve – issues raised by a globally burgeoning industrial reserve, see Brass (2023d). On the earlier views of Beveridge, see below.

By contrast, those who currently do not address what Marx said about the industrial reserve, subscribe instead to an epistemologically and politically different interpretation. Unlike Marxists, whose materialist analyses have been – and are still – on the disempowerment of class, exponents of 'new' populist postmodern theory associated with the 'cultural turn' focus on the empowerment of non-class identity (ethnicity, nationality). Consequently, for them the process of migration is to be defined in positive terms, as the exercise by the 'subaltern'/'multitude' from Third World nations of its human rights. Thus access to the labour markets of metropolitan capitalist nations is depicted as no more than equitable reparations for the long history of exploitation/ oppression their populations experienced as a result of colonization.

The absence of references to the industrial reserve, or downplaying the analytical importance of its current role in and impact on accumulation, together with its form of class struggle, has to be seen as part of a much wider political and intellectual process. On the one hand the expulsion from the Marxist canon of key concepts, all central to its logic, while on the other attempts at inclusion within the same theoretical framework of non- and even anti-Marxist constructs, all of which negate the Marxist political dynamic. Over the recent past it has been possible to encounter claims to be following a Marxist approach by those who have in effect stripped away any or all of its main conceptual apparatus.[4] What one is left with in these circumstances is, variously, Marxism-without-value-theory, Marxism-without-a-proletariat, Marxism-without-class, Marxism-without-struggle, Marxism-without-revolution, and Marxism-without-a-socialist-transition.[5]

To this list can be added two further variants. First, Marxism-without-the-industrial-reserve: and second, the-industrial-reserve-as-a-form-of-worker-empowerment, in effect turning Marx on his head: a similar negation, albeit of a different kind. Those going down these two paths, each of which is the subject of the analysis which follows, usually signal this fact with announcements of an intent to reinterpret Marxism, on the basis not of what Marx actually wrote but rather what he really meant, what he might have meant had he thought about it some more, and what he would undoubtedly have said or written had he not at the time fallen asleep in an armchair in front of the fire.

---

4   In the course of the last two decades, a number of book chapters, critical articles, review articles, and book reviews by me have chartered both the fact of and the reasons for this kind of shift.
5   Almost a century ago, similar kinds of approach were labelled by Trotsky (1934) as accounts of the 1917 Russian Revolution written by epigones, or 'disciples who corrupt the doctrines of their teachers'.

This chapter is composed of two sections, the first of which examines why historically both Marxism and liberalism have regarded the industrial reserve as negative. By contrast, the second considers why more recent interpretations have redefined surplus labour as positive. It is argued here that, as long as capital has access to the industrial reserve, possessing thereby a capacity to restructure its labour process, this will continue to pose difficulties for the formation/consolidation by *all* workforce components of a uniform political consciousness and organizational solidarity.

I

Marxists generally, and Marx together with Engels in particular, have always been clear about the interrelatedness of the formation, the characteristics and the crucial political and economic role of the industrial reserve. Unlike other approaches, which perceived surplus labour as an unintended development (= anomaly), Marxist theory categorized the industrial reserve as the *sine qua non* of the accumulation process, a major weapon in the class struggle available to producers. Consisting of those who are either under-employed or wholly unemployed, the industrial reserve ensures that wages and conditions will always remain at levels below what they might otherwise reach.[6] Though not absolute, as Marx himself accepted, the general impoverishment (= 'immizeration') of the workforce as a whole means that the 'more extensive ... the pauperized sections of the working class and the industrial reserve army, the greater is official pauperism'. This, emphasized Marx, '*is the absolute general law of capitalist accumulation*'.[7]

19th Century Marxist Views

The industrial reserve operates in seemingly contradictory ways. On the one hand, it holds some elements that producers may never call upon as full-time replacement employees: as well as being a source of labour-power that is unfree, the category of surplus labour also contains those whom Marx described as

---

6 On this point, Marx (1976: 790) observed: 'Taking them as a whole, the general movements of wages are exclusively regulated by the expansion and contraction of the industrial reserve army ... [t]hey are not therefore determined by the variations of the absolute numbers of the working population, but by the varying proportions in which the working class is divided into an active army and a reserve army, by the increase or diminution in the relative amount of the surplus population, by the extent to which it is alternately absorbed and set free'.

7 Marx (1976: 798, original emphasis).

belonging to the lumpenproletariat (vagabonds, criminals, prostitutes).[8] Despite remaining outside the labour market, such components remain valuable for capital as a potential threat – replacement workers, strike-breakers – to permanent labour that is well-paid. On the other, the rapid use-up of labour-power means that capital needs access to the industrial reserve merely to replenish its workforce, a point Marx stressed.[9] Much the same emphasis was made by Engels when observing both that 'the length of life of labour-power is immaterial to the capitalists', and that '[t]he capitalist sees only the continuously available surplus population and wears it out'.[10] The latter notwithstanding, Marx was adamant that the reproduction of the industrial reserve was not determined by population increase.[11]

These contrasting economic roles – those constantly required as replacement labour, those never to be employed as replacement workers – of the industrial reserve combine to regulate the labour market on behalf of capital.[12] Hence the presence of the industrial reserve serves multiple ends: it keeps wages down and discourages the formation by all workers of an inclusive consciousness of class.[13] The significance of the latter objective is crucial: the

---

8   See Marx (1976: 797), who elaborates: 'Pauperism is the hospital of the active labour-army and the dead weight of the industrial reserve army'.
9   According to Marx (1976: 795), therefore, 'the consumption of labour-power by capital is so rapid that the worker has already more or less completely lived himself out when he is only half-way through his life [and] falls into the ranks of the surplus population'.
10 Having calculated the level of exploitation inherent in the working day, Engels (n.d.-b: 69, original emphasis) commented: 'These facts prove that capital regards the labourer as nothing else than *labour-power*, all of whose time is labour-time to the extent that this itself is at all possible at a given moment, and that the length of life of labour-power is immaterial to the capitalist [who] sees only the continuously available surplus population and wears it out … Capital is ruthless towards the health and length of life of the labourer'.
11 'Capitalist production can by no means content itself with the quantity of disposable labour-power which the natural increase of population yields', noted Marx (1976: 788), since '[i]t requires for its unrestricted activity and industrial reserve army which is independent of these natural limits'.
12 As Marx (1976: 792) notes: 'The industrial reserve army, during periods of stagnation and average prosperity, weighs down the active army of workers; during periods of overproduction and feverish activity, it puts a curb on their pretensions. The relative surplus population is therefore the background against which the law of the demand and supply of labour does its work. It confines the field of action of this law to the limits absolutely convenient to capital's drive to exploit and dominate the workers'.
13 Among those who recognized this was Engels, who in a letter to Schlüter, dated 30 March 1892, observed (Marx and Engels, 1934: 496–7): 'Your great obstacle in America, it seems to me, lies in the exceptional position of the native workers. Up to 1848 one could only speak of a permanent native working class as an exception: the small beginnings of it in

industrial reserve hinders long-term goals, and how common political interests might be realized politically as a result of organization in pursuit not just of improved conditions and higher wages but also of broader systemic change, designed to bring about a socialist transition. Instead of the latter, the industrial reserve generates splits within the ranks of labour together with the privileging by workers of non-class identities as a way of protecting existing jobs. Moreover, this is a reactionary trend that becomes more important with the global spread of capitalism, the consequent internationalisation of labour market competition, and the increased levels of immigration.

As pointed out by Engels, the introduction of machinery into the industrial labour process not only enhances productive efficiency but also throws more and more people out of work who then – as members of the industrial reserve army – can be used by capital to force down the wages/conditions of those who remain in employment.[14] Current forms and pace of deskilling, plus the

---

the cities in the East always had still the hope of becoming farmers or bourgeois. Now a working class has developed and has also to a great extent organized itself on trade union lines ... [However,] immigrants are divided into different nationalities and understand neither one another nor, for the most part, the language of the country. And your bourgeoisie knows much better even than the Austrian Government how to play off one nationality against the other: Jews, Italians, Bohemians, etc., against Germans and Irish, and each one against the other, so that differences in the standard of life of different workers exist, I believe, in New York to an extent unheard of elsewhere. And added to this is the total indifference of a society which has grown up on a purely capitalist basis ... towards the human lives which succumb in the competitive struggle: "there will be plenty more, and more than we want, of these damned Dutchmen, Irishmen, Italians, Jews and Hungarians;" and beyond them in the background stands John Chinaman, who far surpasses them all in his ability to live on dirt'.

14   'If the introduction and increase of machinery meant the displacement of millions of hand workers by a few machine workers', Engels (n.d.- a: 307–8) noted, 'the improvement of machinery means the displacement of larger and larger numbers of machine workers themselves, and ultimately the creation of a mass of available wage workers exceeding the average requirements of capital for labour – a complete industrial reserve army, as I called it as long ago as 1845 [in *The Condition of the Working Class in England*] – a reserve that would be available at periods when industry was working at high pressure, but would be thrown out onto the streets by the crash inevitably following the boom; a reserve that would at all times be like a leaden weight on the feet of the working class in their fight for existence against capital, a regulator to keep wages down to the low level which suits the needs of capital. Thus it comes about that machinery, to use Marx's phrase, becomes the most powerful weapon in the war of capital against the working class, that the instruments of labour constantly tear the means of subsistence out of the hands of the labourer, that the very product of the labourer is turned into an instrument for his subjection ... Thus it comes about that the excessive labour of some becomes the necessary condition for the lack of employment of others, and that large-scale industry, which hunts all over

enhanced rapidity of labour-power use-up, make possible and indeed necessary the activation of hitherto unutilized elements belonging to the industrial reserve army of labour.[15] Although the latter encompasses those thrown out of work as a result of mechanization and technification, members of the reserve army can be incorporated by capital into its labour process for two distinct reasons. The first refers to a reserve to be drawn on when market demand expands, and capitalists need further amounts of labour-power. By contrast, the second refers to those who – as a mass of unemployed also part of the reserve army – are drawn on by capitalists not so much to increase production but rather as a weapon in their struggle against those still in work.[16] In this second role, members of the reserve army are no longer used simply in addition to an existing workforce but now instead of the latter.

Accordingly, it is crucial to distinguish between two kinds of migrant labour: that bought in *to supplement* the existing workforce, because the latter is insufficient to meet the needs of production. This is termed an as-well-as arrangement. The second form possesses a very different dynamic: migrants recruited in order *to displace* the existing workforce, because the latter either won't work for the low pay and conditions on offer, or – if in post – are deemed too costly to employ, the object being to replace it with cheaper foreign labour. This is termed an instead-of arrangement, one that historically and currently generates huge antagonism within the ranks of the working class affected, an hostility that frequently resorts to discourse about ethnic, national or gender 'otherness'.

---

the world for new consumers, restricts the consumption of the masses at home to famine minimum and thereby undermines its own internal market'.

15 On the rapidity of labour-power use-up, Marx (1976: 789–90) commented: 'The over-work of the employed part of the working class swells the ranks of its reserve, while, conversely, the greater pressure that the reserve by its competition exerts on the employed workers forces them to submit to over-work and subjects them to the dictates of capital. The condemnation of one part of the working class to enforced idleness by the over-work of the other part, and *vice-versa*, becomes a means of enriching the individual capitalists, and accelerates at the same time the production of the industrial reserve army on a scale corresponding with the progress of social accumulation'.

16 Sweezy (1946: 99, emphasis added) recognized the importance of this distinction, observing that 'the increasing use of machinery, which in itself means a higher organic composition of capital, sets free workers and thus creates "relative overpopulation" or the reserve army. Marx stresses the point that the existence of unemployed labourers is conducive to the setting up of new industries with a relatively low organic composition of capital and hence a relatively high rate of profit 'It would seem, however, that a more important effect of the reserve army is ... through competition on the labour market with the active labour force, to depress the rate of wages and in this way to elevate the rate of surplus value'.

It is argued here that the combination of the global spread of capitalist development, the internationalization of the industrial reserve army, and the restructuring of the labour process, has been accompanied by the instead-of form. Significantly, perhaps, this does not prevent employers from insisting that they recruit migrants only because no locals are available or willing to do the work, a thinly disguised attempt to represent merely as supplementing what is actually its 'other' – displacing the existing workforce. Clearly, it is the latter form that drives capitalist restructuring, given the importance of cost considerations where accumulation is concerned. Unsurprisingly, therefore, when asked why they prefer a migrant workforce, capitalists deploy the politically less contentious addition-to version rather than the instead-of form.

The assumption that those expelled from the labour force by the application of machinery and technology would – after a brief sojourn in the reserve army – find alternative employment, is both pervasive and misplaced. It is linked to the notion that such workers would be employed as a result of investment in new enterprises by capitalists whose enhanced profits derived from the original labour-displacing strategy. This view is problematic, since it applies only when capital and labour are national in scope, and not international. Where the latter is the case, capital is able to do two things: either to invest elsewhere, in contexts where labour-power is available and even cheaper; or to employ (perhaps even to import) migrants who meet the same requirements. Whichever the instance, the outcome is the same: any new jobs created do not necessarily go to those expelled from the labour process because of technification/mechanization of production. The obviousness of this outcome notwithstanding, it is unfortunately still possible to hear – even from some on the left – the mantra that workers displaced in this manner will automatically find employment in new industries created by capital.

In the course of the Great Depression of the 1930s, the 1939–45 war, and its immediate aftermath, when Keynesianism was said to have solved the issue of unemployment, the industrial reserve army again moved to the centre of the political agenda, both for the liberalism of Beveridge and for neo-Marxism of Kalecki, and the Marxism of Sweezy and Dobb at that early conjuncture, and – much later – of Glyn. Notwithstanding their political differences, the negative perception both that the connection between unemployment and the continued access by capital to unregulated forms/sources of surplus labour, and that state intervention was required in order to break this connection, was an interpretation shared by all of them.

20th Century Liberal Views

Liberal ambivalence concerning the role, impact, and desirability of the industrial reserve structured the approach during the 1930s and 1940s of William Beveridge to the creation in post-war Britain of the welfare state. Anxiety regarding what he termed the 'glutting of the labour market', linked by him both to 'under-employment' as a form of sweating, and to the beating down of average earnings to subsistence level, was balanced against the need as he saw it of how to provide properly maintained reserves of labour. Underwriting this ambivalence was apprehension that a failure to address the economic connection between these phenomena might undermine the legitimacy of the accumulation process itself, and thus empower those advocating socialism. Beveridge himself recognized the possibility of such a link, conceding that 'the most general effect of war is to make the common people more important'.[17]

Of significance, therefore, is that Beveridge traced the perpetuation of what he termed 'the four giant evils' – Disease, Ignorance, Squalor, and Idleness – back to the presence and operation during the pre-war era of the reserve army of labour.[18] His 'perplexity' about the latter stems from what Beveridge regards as the 'central paradox of the unemployed problem': the juxtaposition between on the one hand rising remuneration of labour and on the other the 'irreducible' growth of unemployment.[19] Although as a liberal he appeared mystified

---

17 Beveridge (1943: 109). Even before the 1939–45 war, Beveridge (International Labour Office, 1924: 7) drew attention not just to the long-recognized connection between casual employment, labour market competition, economic crisis and increasing poverty, but also to the necessity of addressing this politically. Hence the observation that '[i]t has been for years now established how casual employment produces unemployment, and the remedy by what is called decasualisation is equally certain and agreed. It is little short of a scandal that, merely because of the difficulty of getting things done practically, we leave that side of the problem of unemployment practically untouched'.

18 About the link between the industrial reserve, unemployment, and impoverishment, Beveridge (1931a: 107–108) observes: 'The social consequences of this under-employment and of under-payment or sweating … are ultimately indistinguishable. Each means the maintenance, as an integral part of industry, of a low and miserable form of life … Here if anywhere is to be seen the beating down of the remuneration of labour under competition to bare subsistence … By casual employment therefore real earnings may be and are driven down to a normal level far below the lowest rate possible in regular industry however plentiful the competition and unorganized the workmen'.

19 See Beveridge (1931a: 70), who concludes (Beveridge, 1931a: 95) that '[t]he system of casual employment to meet fluctuations [in business] requires the maintenance of reserves of labour at all the points at which men are engaged. A considerable part of these reserves is so irregularly employed as to be in chronic poverty. On every day some part of them is standing idle'. This describes as clearly as necessary one major economic role of the industrial reserve: merely to 'be there', to be called upon as and when necessary.

by this, a Marxist would point out that it corresponds to a situation in which the industrial reserve comes into its own: when faced with rising wages (a tight labour market) employers resort to cheaper forms of labour-power.[20] Hence not merely to the liking of capitalist producers, but crucially their need for the industrial reserve, plentiful and on station, to be drawn upon as and when required.[21]

As significant is that his motivation for advocating the welfare state as a solution to the impoverishment and misery inflicted by the industrial reserve during 1930s Depression, therefore, was in part due to a wish to avoid a 'from below' challenge to any attempt at a return to pre-war economic and social conditions.[22] This concern was itself signalled in the epigraph on the title page of the Beveridge report: 'Misery generates hate', not just between different components of the working population but also between the latter and employers.[23] An additional political concern was that a failure to ameliorate the impact of the industrial reserve would fuel the rise of racism. Beveridge saw mass unemployment as leading inevitably to a war of all against all, as

---

20   Commenting on those employed in unskilled and unorganized work, Beveridge (1931a: 69–70) noted: 'The glut of labour in them is notorious. Has there ever ... been a time when employers could not get practically at a moment's notice all the labourers they required? Is not this indeed the root of bewilderment and despair in regard to the unemployed problem that there appears to be always and everywhere an inexhaustible excess in the supply of labour over the demand?'.

21   Beveridge (1931a: 76, 80) outlines the manner in which producers operate their labour reserve: 'The general formula for the supply of labour in an industry appears then to be this: for work requiring ... at most ninety-eight men, there will actually be eighty in regular employment and twenty in irregular employment; there will be a hundred in all ... [t]he twenty, however, are as much part of the industrial system as are the eighty; the reserve is as indispensable as the regulars ... every element of chance in the competition for employment ... tends to swell the actual number of individuals between whom any definite amount of work is distributed and to decrease the share of each, down to the limit fixed by the standard of subsistence'.

22   'I want to be quite certain than I can change the person who governs me without having to shoot him', argued Beveridge (1943: 92), adding: 'That is the essence of Democracy, that you can have a peaceful change of governors without shooting. To me a country is not a Democracy ... if you cannot change the Government by a perfectly peaceful method of putting your cross on a piece of paper'.

23   The epigraph, Beveridge (1944: 15–16) explained, 'comes from the account given by Charlotte Brontë, in the second chapter of *Shirley*, of the handloom weavers who one hundred and twenty-five years ago were being driven into unemployment and miserable revolt by the introduction of knitting frames'. He continues: 'To look to individual employers for the maintenance of demand and full employment is absurd. These things are not within the power of employers. They must therefore be undertaken by the State, under the supervision and pressure of democracy, applied through the Parliament men'.

every person 'appears as the enemy of his fellows in the scramble for jobs'.[24] The latter in turn generates 'the growth of jealous restrictions', including controls on the free movement of labour and opposition to technology, fostering 'still uglier growths' such as anti-semitism, anti-foreigner sentiment, and hostility to the employment of women as workers. All these developments were attributed by him to a 'failure to use our productive powers' in order to solve unemployment: only when this had been done would there no longer be conflict between capital and labour.

In the opinion of Beveridge, however, the goal of full employment was not just unrealizable but undesirable. Consequently, the industrial reserve will never entirely disappear, nor did he think it desirable that it should do so. This he attributed both to seasonal fluctuations in demand for labour, and to an insistence on his part that the policies advocated 'does not mean giving to everyone security in his particular job'.[25] As a political liberal, therefore, Beveridge still viewed the market as making a positive contribution to society. Because he believed in 'citizen liberties' and private enterprise – 'on condition that those liberties are exercised responsibly' – Beveridge joined the Liberal Party.[26] Whereas the economic programme of the Labour Party at that conjuncture was premised on nationalisation of key industries, so as to enable both regulation and planning, Beveridge had 'no bias in favour of nationalisation' and was content to leave 'much the greater part of industry to private enterprise'.[27]

In keeping with Keynesian demand management, what the Beveridge Report advocated was using the State to give workers more purchasing power,

---

24 'So long as chronic mass unemployment seems possible', noted Beveridge (1944: 248), 'each man appears as the enemy of fellows in the scramble for jobs. As long as there is a scramble for jobs it is idle to deplore the inevitable growth of jealous restrictions, of demarcations, of organized or voluntary limitations of output, of resistance to technical advance. By this scramble are fostered many still uglier growths – hatred of foreigners, hatred of Jews, enmity between the sexes. Failure to use our productive powers is the source of an interminable succession of evils. When that failure has been overcome, the way will be open to progress in unity without fear'.
25 Beveridge (1944: 126ff.).
26 Beveridge (1946: 57).
27 Beveridge (1946: 63). His commitment to the continuation of the market is evident from an earlier defence of *laissez-faire* economic policy, where – along with co-authors to an edited volume on tariffs (Beveridge, 1931b: VI) – it was stated that 'we should all think it a disaster, if the policy of Free Trade which has served Britain so well materially, as through her it has served as an inspiration to all who in any land have worked for a good understanding among nations, were today to be sacrificed to ignorance or panic or jealousy or specious calculations of a moment's gain'.

the spending of which would in turn create demand for commodities manufactured by capitalist producers in Britain.[28] Its object was to establish some control over an unplanned market economy, not its elimination: in the words of Beveridge, 'planned marketing and production of primary products, both agricultural and mineral, is an essential condition for the stabilization of demand for manufacturing products'.[29] His aim was to save capitalism, and he obtained the support of conservatives and employers for his programme by arguing not just that better-paid workers formed consumers for commodities but also that healthcare and other welfare costs would henceforth be met by the state.

20th Century Marxist Views

During the 1940s and 1950s, amidst end-of-ideology triumphalism accompanying claims that Keynesianism had banished recurring capitalist crises, perceived as the ability of capital actually/potentially to produce abundance, thereby providing workers with full employment and higher living standards, Marxists took a different view. Unlike liberals such as Beveridge, who thought that full employment was unachievable, Marxist/neo-Marxist economists like Kalecki, Sweezy, and Dobb argued that full employment was never going to be acceptable to capitalists. Each of them positioned the issue of the industrial reserve in relation to the need on the part of producers for enhanced labour market competition in order to impose a check on wages and boost profitability, and thus untrammelled access to the industrial reserve. The difficulty faced by producers is simply put: where capitalist demand for labour-power exceeds supply, wages tend to rise. What, therefore, asks Sweezy, 'keeps wages in check so that surplus value and accumulation may continue as characteristic and essential features of capitalist production?'.[30]

The answer to this, Sweezy indicated, was the presence of the reserve army, which 'consists of unemployed workers who, through their active competition

---

28　On the necessity of a greater role for the state in post-war Britain, see Beveridge (1943: 90ff.). His support for Keynesian theory is also clearly signalled (Beveridge, 1944: 106–7).

29　Beveridge (1944: 103). Significantly, in the 1945 British general election – which Labour won by a landslide – the programme of the Conservative Party allocated economic importance to the continued role of the market ('free enterprise'), whereas that of the Labour Party sought to bring about full employment in industry by means of state planning (McCallum & Readman, 1947: 53–54).

30　Sweezy (1942: 87).

on the labour market, exercise a continuous downward pressure on the wage level'. This was linked by him to the problem of crisis, which stems from a situation where the reserve army, for whatever reason, shrinks, thereby eliminating the check it exercises on wages.[31] The result is that '[c]apitalists are forced to bid against one another for additional workers, wages rise, and surplus value is cut into', with negative outcomes for investment and accumulation.[32] When wages start to increase, then is when the real significance of the industrial reserve comes into its own, as can be seen currently in cases where employers and their organizations call for more access to foreign workers: this is because 'the stronger the tendency of wages to rise, the stronger also will be the counteracting pressure of the reserve army'.[33]

In a lecture at Cambridge during 1942, Kalecki highlighted the potential fact of employer opposition to contraction in the industrial reserve on the grounds that 'lasting full employment is unsound from their point of view and that unemployment is an integral part of the "normal" capitalist system'.[34] For his part, Sweezy – like Marx – reiterated the point that the industrial reserve operates independently of population levels, underlining that 'Marx's great accomplishment was … to free [the conceptualisation of the industrial reserve] from an otherwise fatal dependence on the Malthusian population dogma'.[35] At the start of the following decade, Dobb reinforced the earlier concern expressed by Kalecki, warning against the illusion that accumulation and full employment

---

31   '[T]hrough its relation to the reserve army', explained Sweezy (1942: 90), 'the problem of crises assumes a central position in Marx's theoretical system. Whereas for the classical theorists, the problem was not so much to explain crises as to explain them away, for Marx capitalism without crises would be, in the final analysis, inconceivable'.

32   Sweezy (1942: 150).

33   Sweezy (1942: 88).

34   'We have considered the political reasons for the opposition to the policy of creating employment by government spending', noted Kalecki (1943: 326), adding: 'But even if this opposition were overcome – as it may well be under the pressure of the masses – the maintenance of full employment would cause social and political changes which would give new impetus to the opposition of business leaders. Indeed, under a regime of permanent full employment … [t]he social position of the boss would be undermined, and the self-assurance and class-consciousness of the working class would grow. Strikes for wage increases and improvements in conditions of work would create political tension … [The] class instinct [of business leaders] tells them that lasting full employment is unsound from their point of view and that unemployment is an integral part of the "normal" capitalist system'.

35   Sweezy (1942: 89).

were compatible.[36] Presciently, it was pointed out by him that, sooner or later, capitalist profitability would require the reintroduction of unemployment, and with it an enhanced industrial reserve, and this is exactly what occurred, culminating in the rise and consolidation of *laissez-faire* deregulation from the 1980s onwards.

Maintaining or enhancing profitability in this way is crucial for any accumulation project in what is an increasingly competitive world market. For this reason, the global spread of capital during the post-war era made access to the industrial reserve not just easier but necessary. From the 1960s onwards the Green Revolution drove peasants in Third World nations (India, Mexico) off the land, thereby commencing the augmention of the industrial reserve world-wide, while from the 1990s onwards the fall of the USSR released new sources of labour from Eastern Europe available for employment by capital in the European Union. Combined with the latter is the outsourcing made possible by the expansion in the industrial reserve throughout Asia, not least that of China, a development licensing what Glyn has referred to as a contraction in labour's share of national income coupled with 'Marx's rising rate of exploitation emerging, a century and a half after he first predicted it'.[37]

It has therefore become possible for corporations either to outsource production to far-off locations where surplus labour already existed, or else to insource labour itself, from these same and other locations closer to where production was already based. Drawing on either or both these supplies of worker meant that producers were henceforth able to compete with rival capitalists as long as they restructured their own labour process. This employers did by

---

36 See Dobb (1955: 215–25), where he argued that capitalists viewed full employment as 'a situation where the sack has lost a good deal of its sting as a disciplinary weapon, with the virtual disappearance of the industrial reserve army'.

37 See Glyn (2006: 14), who comments: 'What makes China (and India) fundamentally different is the presence of vast reserves of labour previously isolated from the world economy ... This represents an enormous potential labour supply. Estimates of the numbers who may be pulled out of agriculture, where their incomes are very low, into industrial and service jobs in the the towns range as high as 150–300 million ... These, together with tens of millions of urban unemployed, constitute a reserve army of labour of quite unprecedented magnitude'. He continues (Glyn, 2006: 15): 'China is producing large numbers of highly trained but still relatively cheap workers ... This could bring intense pressure on the jobs and working conditions of northern workers [as the] bargaining chips would be in the hands of capital to a degree not seen since the industrial revolution. The stylised fact of labour's share fluctuating in the 2/3 to 4/5 range could disappear too, with Marx's rising rate of exploitation emerging, a century and a half after he first predicted it'. Further details about the post-war background leading to this process can be found in Glyn (2007).

replacing those in better-paid permanent jobs with cheaper workers, either at home or abroad. Those belonging to the latter category are employed usually on a temporary or casual basis, a transformation that in some instances also corresponds to multiple forms of workforce substitution: of foreign migrants for locals, of unfree production relations for pre-existing free equivalents, and of age- or gender-specific forms of labour.[38]

Until the 1980s this restructuring was not possible: in 1960s Italy, for example, the strength of working class organization and struggle, plus the absence as yet of cheap migrant labour, combined to prevent producers from taking advantage of the industrial reserve. 'The main point', noted one observer, 'is whether or not secondary workers are available to act as a potential source of competition for primary workers'.[39] The neoliberal turn from the 1980s, however, involving as it did deregulation, privatisation, and the adoption of *laissez-faire* economic policies, facilitated the cross-border movement of workers and money: the transfer of capital and labour between sending and receiving countries paved the way ultimately for segmentation of the labour market within metropolitan capitalist nations themselves. Such labour process restructuring, involving workforce decomposition/recomposition, was anticipated by Marx, and is described by him thus: 'On the one hand, therefore, with the progress of accumulation a larger variable capital sets more labour in motion without enlisting more workers; on the other, a variable capital of the same magnitude sets in motion more labour with the same mass of labour-power; and, finally, a

---

38 In the words of Marx (1976: 788), 'the development of the capitalist mode of production, and of the productivity of labour ... enables the capitalist, with the same outlay of variable capital, to set in motion more labour by greater exploitation (extensive or intensive) of each individual labour-power. [Hence] the capitalist buys with the same capital a greater mass of labour-power, as he progressively replaces skilled workers by less skilled, mature labour-power by immature, male by female, that of adults by that of young persons or children'. The fact of and the importance of changes in the age and gender division of labour, together with the link to the industrial reserve, are issues emphasized subsequently (Marx, 1976: 794–95): 'Both in the factories proper, and in the larger workshops ... large numbers of male workers are employed up to the age of maturity, but not beyond. Once they reach maturity, only a very small number continue to find employment in the same branches of industry, while the majority are regularly dismissed. This majority forms an element of the floating surplus population which grows with the extension of those branches of industry ... Capital demands more youthful workers, fewer adults'.

39 See Bruno (1979: 131) who points out that 'in Italy this competition did not take place, not only because the working class employed in the core of large and medium industrial firms (primary workers) resisted it successfully, but also because the surplus population (secondary workers) did not compete in the primary market'.

greater number of inferior labour-powers is set in motion by the displacement of more skilled labour-powers'.[40]

II

In a basic sense, the second part of this presentation marks what can only be described as a fundamental epistemological and political shift in the way the industrial reserve is interpreted. This entails a change from the undeniably negative one, held both by Marxism and by some liberals, to views – held by Basso, Mezzadra, and Bradley and Norhona, among others – that regard the industrial reserve as in some way positive.[41] The resulting break extends from those who see the industrial reserve merely as unimportant, or else politically empowering for the working class as a whole, as evidence for its progressive role in challenging capitalism, to those with a different focus, who regard it simply as empowering for migrants.[42] A variant of this is Basso, who insists that over the recent past migrant workers cannot be said to form part of the industrial reserve, since 'it remains the case that the majority of immigrant workers in the EU *are within the active army of labour, not the reserve army*'.[43] Thus the element of surplus labour has not so much vanished from the development debate as been redefined.

This negative→positive shift accompanied and indeed was made possible by the wider context: the rise of the 'new' populist postmodernism (= the 'cultural turn') which marked a privileging of non-class identity (ethnicity, nationality) and the conceptual displacement of Marxist theory about class and class struggle. It was this as much as anything that led to a move away from the industrial

---

40 Marx (1976: 788). However, the many references by Marx (and other Marxists) to the important role of female and child labour-power in the way producers have restructured the labour process historically, and still do so today, has not prevented some current observers from continuing to claim – erroneously – that Marxism ignores the fact that not all workers employed by capital are adult males.

41 Considered elsewhere (Brass, 2017b: Chapter 19; Brass, 2021b: Chapter 8; Brass, 2022a: Chapter 7) are two other views about the industrial reserve, neither of which name it as such. On the one hand, therefore, are bourgeois economists who, like employers, regard surplus labour as a positive contribution to economic growth. On the other is the negative interpretation held by exponents of Great Replacement theory, which regards immigration simply as a process of cultural erosion.

42 Those who dismiss the industrial reserve as unimportant extend from the 1980s, when 'a number of recent feminist writers ... dismissed the relevance of the industrial reserve army to an understanding of women in capitalist society' (on which see Collins, 1984: 52) to its present categorization as 'functionalist' by an agrarian populist (Bernstein, 2021b: 26, n 21). A variant of this view simply denies that surplus labour is a political issue, for an instance of which see 'The Tories concocted the myth of the "migrant crisis"', *The Guardian* (London), 7th November 2022.

43 Basso (2021: 6–7, original emphasis).

reserve as a locus/cause of worker disempowerment and to its redefinition as the empowerment of 'otherness', to be accepted and celebrated as such.[44] The latter was signalled by the way in which migration into metropolitan capitalist nations now became characterized in positive terms, as a 'human right' reflecting a 'common humanity', no more than a form of reparation owed by receiving countries to sending ones for having colonized them.

Border Wars

To this category it is necessary to add yet another view, drawn significantly from within Marxist theory: a positive one that claims to reinterpret what Marx himself meant, and on the basis of this to argue that the industrial reserve is the locus of struggle to undermine not working class solidarity but much rather capitalism itself. Unsurprisingly, given a shared positive take on migration into metropolitan capitalism, this championing of an open-door policy in the name of Marx combines with similar calls made in the name of non-class identity. Accordingly, opposition to racism within the nation state – an uncontroversial proposition backed by all Marxists – is interpreted by exponents of 'the cultural turn' as an endorsement of the unconditional right to cross-border migration simply on the grounds of being 'other'. This in turn licenses an imperceptible slide into ideological support not just for open-door policy but also and thereby for an expanding industrial reserve, which fuels the political emergence of rival populisms. To the postmodern argument emphasizing the cultural identity of the migrant-as-'other'-nationality, therefore, the far right counterposes an argument similarly emphasizing cultural identity, only this time the nationality of the non-migrant worker.

According to Basso, Marx is wrongly invoked in order to justify anti-immigrant border controls, that it is his ideas about the capitalist industrial reserve that are deployed most frequently in this endeavour (especially in Italy and Germany), and that such interpretations distort Marx and decouples it

---

44 This shift reproduces the earlier approach to the question of immigration of Kennedy (1964), in whose footsteps Basso *et al* seemingly follow. Accepting that immigration was 'not always a happy experience', the issue for Kennedy (1964: 67–68) was one of 'adjustment' and 'assimilation', a process amounting to 'the expression in action of a positive belief in the possibility of a better life'. For current versions of the same argument, see Marquardt (2021) and Khanna (2022), together with an opinion piece by the latter 'Borders are holding back the world's eight billion', *Financial Times* (London), 12–13 November 2022.

from his political conclusions.[45] In the opinion of Basso, therefore, 'those who have claimed to be drawing on Marx's analysis of the industrial reserve army ... have distorted it and severed it from his political conclusions'.[46] For this reason, opponents of permanently open-door immigration into the EU who base this on what Marx wrote about the industrial reserve are nevertheless described by Basso as Nazis and racists, a label applied to three commentators in particular: Diego Fusaro, Sahra Wagenknecht, and Wolfgang Streek.[47] A consequence of their breaking with Marxist internationalism, argues Basso, is an espousal of nationalism. According to each of them, therefore, immigration supresses national wage levels, national resources are insufficient to meet the requirements of both locals and immigrants, and immigration generates culture wars within the nation.[48]

Basso is right to blame Fusaro for thinking that the capitalist state can be relied on to regulate the labour market on behalf of workers, a view Basso correctly describes as 'empty idealism'.[49] Marx himself certainly did not believe this to be the case, although currently there are indeed purportedly leftist analyses in the sphere of development studies which maintain just this, arguing for political solutions to *laissez-faire* that simply entail a return to a 'kinder'/'caring' capitalist state.[50] For this reason Basso is also right to criticize the nationalism not just of Fusaro but also of Wagenknecht and Streek, all of whom imagine that working class political interests can safely be left in the hands of a state apparatus that is capitalist.

In defence of their core argument – that borders should be abolished – Bradley and Noronha take issue with the leftist case that such a policy would undermine both the livelihood and the organizational capacity of workers

---

45   See Basso (2021: 217–238). The references here are to a copy of the same text but with a different numbering (Basso, 2021: 1–21), kindly sent me by its author.
46   Basso (2021: 8ff.).
47   Wagenknecht is a member of the left in the Bundestag and author of a critique of identity politics (*Die Selbstgerechten*/The self-righteous), while Streek (2017) is a sociologist who has written about the political economy of German society. Each is accused (Basso, 2021: 13) of 'falsification of Marx's thought', and further 'the posture of both our heroes is of people who stand firmly on the side of the weakest, of the "lower classes", the working class, against the strong powers of global capital that inordinately swell the industrial army with migrants'.
48   For his part, Streek wishes to close borders to asylum seekers as well as economic migrants.
49   Basso (2021: 10–12).
50   Among those who advocate a return to a 'kinder'/'caring' capitalist state is Jan Breman (on which see Brass 2018a: Chapter 5).

currently in jobs.[51] It is clear, moreover, that the migration pattern they envisage is one composed of those who want the right 'to work, to join family, to access welfare benefits and healthcare, and to move freely'.[52] Border abolition and its 'struggle for freedom', they then observe, licenses nothing less than an ability 'to move and stay', which is precisely the kind of policy favoured by capitalists.[53] Subsequently they seem to recognize that this case about the desirability of open borders is in lockstep with the pursuit by capital of access to the industrial reserve: this notwithstanding, they shift blame away from capital and towards colonialism, doubling down on their original argument.[54]

Equating border controls with a colonial legacy that discriminates against ethnic 'otherness', underlines the extent to which Bradley and Noronha regard any objection to their advocacy of an open-door approach as merely racist.[55] It comes as no surprise that any questioning of – let alone expressing concerns about – immigration is dismissed by them simply as evidence of racism.[56] In what is a reductive approach, therefore, everything is presented as an effect of

---

51 Noting that 'The intensification of violent and spectacular bordering is intimately connected to the ascendancy of racist, nationalist and rightwing governments', Bradley and Noronha (2022: 2, 3) continue: 'But this is not only a problem on the right. Voices across the political spectrum assert that borders are sensible and necessary. Many political parties and even trade unions argue that borders protect the working classes from low wages caused by a surplus of migrant labour', before switching to a version of discourse about 'revenge colonialism': 'To sustain this account of borders … requires a deep historical amnesia about colonialism'.

52 See Bradley and Noronha (2022: 6).

53 Bradley and Noronha (2022: 9–10).

54 'Clearly, border abolition needs to be distinguished from … arguments for "open borders"', note Bradley and Noronha (2022: 52), 'and this requires a more critical account of the relationship between capitalism and immigration control'. Much the same contradictory procedure informs their argument regarding nationalism and racism, momentarily conceded as not simply an ideology confined to whites (Bradley and Noronha, 2022: 25–26). Although the fact of ethnic conflict in African and Asian countries after the end of colonial rule is acknowledged, this is attributed largely to the impact of newly independent nations of colonialism.

55 'Despite many successful struggles for formal rights and recognition', insist Bradley and Noronha (2022: 54–55), 'the structural inequalities forged by colonialism remain, mediated more indirectly by immigration controls and restrictive citizenship regimes. This is one of the ways in which we might describe our contemporary world order as racial, or racist: the borders between nation-states perpetuate hierarchies made by colonialism'.

56 See Bradley and Noronha (2022: 15ff., 25, 29), whose catch-all definition of racism is so inclusive that any/every form of economic disadvantage is attributed to this identity. Despite accepting that there are 'obvious class differences' within countries, nationalism is also categorized merely as another kind of racism.

ethnicity, to which the only solution is border abolition ('anti-racists must seek the abolition of borders').

Part of the difficulty faced by Bradley and Noronha is that they conflate two distinct issues, and thus two separate problems. The undifferentiated claim that borders do not 'protect people's rights', therefore, overlooks what under capitalism are divergent interests: of workers in the receiving country, and those in the sending one. There are two ways of looking at borders in a capitalist system, ones that are not just different but potentially antagonistic. To the migrant, a border constitutes a bar to his/her empowerment, in the form of higher wages, improved work conditions, and better livelihood prospects when compared with what exists in the sending country. In the case of those already employed in metropolitan capitalist nations, by contrast, 'rights' – livelihoods, wages/conditions, secure employment that is permanent – are seen as protected by borders, insofar as the latter prevent yet more acute labour market competition.

In keeping with this, Mezzadra looks at the problem of borders simply from the viewpoint of a non-class-specific migrant (= 'the right to escape', 'the autonomy of migration'), not the migrating worker, nor the worker in the country of destination.[57] Such an approach makes it impossible to make common cause with labourers in the latter context when one is effectively in competition with them for the same jobs, the getting of which by the 'escaping' migrant at the expense of an actual/potential worker in another country is the only thing that seems to be seen by Mezzadra as a legitimate political objective.

For his part, Basso argues against '[t]hose who speak of the need to reinstate frontiers, walls and boundaries', which suggests that he, too, favours an open-door approach to the issue of borders.[58] He invokes Marx's internationalism as a justification for an open-door approach, observing that at a time when 'bosses were importing "foreign workers" and "transferring manufacture to countries where there is a cheap labour force"', the struggle against capital 'must become international. Anything but "political control of cross-border flows" and the closing of frontiers!'. Subsequently, however, although this is qualified somewhat, Basso nevertheless insists that eliminating open borders (by regulation/closure) would have no effect on 'workers' problems'.[59]

---

57 Mezzadra (2006).
58 Basso (2021: 11).
59 'As for borders', he comments (Basso, 2021: 16–17), 'it would be naïve to think that their removal would solve everything – that is true. But we can be sure that none of the workers' problems would be solved through the closing of borders, with all the associated rhetoric and racist practices'.

Human Flourishing, but Whose?

The case made by Basso about the industrial reserve is based on a number of problematic claims. To begin with, he doubts that there are those who 'celebrate migration ... as an inherently positive model, at once integrative and emancipatory', an argument which – lacking exponents – is described by him as a 'phantom-subject'. This is quite simply wrong: not only does he himself seemingly endorse just such a positive view (see below), but the list of those who also subscribe to it is long, very long. Supportive of the positive view deemed absent are the non-economic and migrant-centric arguments of much journalistic, NGO, and academic discourse, informed as these are by concepts like justice, citizenship and human rights.[60] The latter also inform the case against borders advanced by Mezzadra and by Bradley and Noronha.

In line with the non-economic epistemology of the 'new' populist postmodernism, justice and citizenship are conceptualized by Mezzadra not in objective but subjective terms: rather than being defined by the receiving country, therefore, citizenship is said to derive simply from grassroots culture of the migrant him/herself.[61] Although misleadingly labelled by him an 'alternative modernity', this notion of citizenship, rooted as it is in a misplaced Thirdworldism, licenses an uncritical acceptance of ideological forms in the sending nation, thereby leaving intact structures advantageous to capitalism. Among the latter are kinship and quasi-kinship relations, the hierarchy and authority of which can be – and have been – used to enforce bonded labour arrangements on migrants.[62]

The object of border abolition, as perceived by Bradley and Noronha, is similarly problematic: 'new ways of caring for one another' leading to 'human flourishing'. In their opinion, 'rights for non-citizens' amounts to 'recuperating

---

60 For details of those holding such views, see Brass (2017b: Chapter 19; and 2021b: 215ff.). More recent examples in the UK press include 'Ukranians could fill job vacancies in Britain, if only they could get visas', *The Guardian* (London), 1st June 2022; and 'Here's the best way for Britain to solve the migrant crisis: give them work visas', *The Guardian* (London), 3rd November 2022. That well-meaning NGOs which supply provisions and/or shelter to those who compose the industrial reserve without addressing the wider systemic cause merely perpetuate its existence was recognized long ago by Beveridge (1931a: 109), who observed that 'the danger of subsidising casual employment by public or private relief without improving the conditions of the casual labourer is a very real one'.

61 See, for example, Mezzadra (2011b), Balibar, Mezzadra and Samaddar (2012), and Mezzadra and Neilson (2013).

62 For the use of kinship and quasi-kinship authority to enforce bonded labour relations, see Brass (1999: 57ff., 125ff.).

the long-unfulfilled promise of human rights'.[63] The opposition by Bradley and Noronha both to citizenship and to nationalism derives from the idea that such identities legitimize and thus underwrite the existence of borders.[64] Unlike Marxism, they interpret immigration control as a relation involving not classes but nations. Since it is this imbalance between 'grossly unequal nation-states' – not classes within them – that gives rise to and sustains borders, Bradley and Noronha maintain that a national boundary together with its accompanying ideology of citizenship perpetuate 'inequality, injustice and harm'. Their case departs from analyses in which the acquisition by the immigrant of citizenship is seen as positive, as a desirable outcome of the migration process: instead, Bradley and Noronha argue that as the concept of citizen is supportive of border divisions, it must be seen as a negative identity, one that justifies barriers separating countries.[65] For them, citizenship rights attached to a metropolitan capitalist nation are exclusionary, aimed at immigrants unable to realize this sort of 'belonging'.

The absence of a sustained consideration of systemic change as seen by political economy, let alone by Marxist theory, does not prevent Bradley and Noronha from claiming that '[b]order abolition is a revolutionary politics' On the issue of borders and immigration, theirs is a maximalist approach: no bar of citizenship acquisition should exist, and anyone from anywhere who wishes to do so must be free to migrate to a metropolitan capitalist nation. It is an interpretation that is problematic on all sorts of levels. To begin with, in economic terms it plays directly into the hands of capital, with its desire to access unlimited sources of the industrial reserve. As important ideologically is that it plays directly into the hands of those on the far right, whose claims about great replacement it appears to vindicate. To choke off the racism informing the latter, it is necessary to prevent yet more labour market competition, which in turn means denying capitalists access to the industrial reserve. Politically, it ignores that even after a socialist transition has been effected, a socialist government retains the power to plan, an executive role that extends to the

---

63 On the centrality of human rights to their approach, see Bradley and Noronha (2022: 28, 32ff., 39).
64 On this, and what follows, see Bradley and Noronha (2022: 4, 28–29) for whom 'anti-racism should centrally include people subject to immigration controls (non-citizens), without trying to resolve the problem by simply turning them into citizens'.
65 'In general', observe Bradley and Noronha (2022: 27–28), 'campaigns for citizenship for particular groups of migrants function to reinforce the notion that you have to be a particular kind of person – a citizen, an insider, someone who belongs – in order to access fundamental rights'.

movement of labour and its allocation under the socialist plan to specific economic tasks in particular areas or locations.

As contentious is the assertion by Basso that all migration is forced, a claim which ignores the distinction made by Marx between labour-power that is free and that which is unfree. Hence the scorn poured on Fusaro for categorizing workers as 'serfs', a disdain expressed by Basso as 'the anti-Marxist one of depicting both immigrants and native-born proletarians as new serfs: the former as mere things, weak, wretched, desperate, vulnerable to blackmail; the latter as beings reified and nullified by the unfortunate arrival of new migrants'.[66] Rather than seeing it in Marxist terms, as a production relation, Basso conflates unfreedom with pre-capitalism (= serf). Consequently, overlooked is the fact that nowadays unfree labour is deployed by capitalists as part of the class struggle waged 'from above'. As such, labour-power that is not free is an important component of the industrial reserve: its acceptability to producers is determined by the very coercion/debt which makes this sort of worker 'vulnerable to blackmail', a relational form which operates against attempts to organize politically and establish common bonds with workers who are free. Bradley and Norhona also misunderstand the significance of unfree labour, and its difference from production relations that are free.[67] Consistent with their privileging of race, what they object to are views which 'define non-citizens solely in terms of their labour power'. Unfree labour is conceptualized, if at all, merely as an epiphenomenon of ethnicity/nationality.

In a similar vein, Mezzadra accepts that his interpretation breaks with Marxist theory. He advocates severing the conceptual link between the wage labour relation and labour-power, so that a 'heterogeneity of labour relations' composed of numerous other social categories and groups (undifferentiated migrants, petty traders, smallholding peasants, family farmers, sharecroppers, lumpenproletarians) can be included among those whose agency will become the deciding factor in any struggle with capitalism.[68] However, since key components of these very broad categories – labelled by him 'subaltern' or 'multitude' – occupy not just different but antagonistic class positions, and thus are usually components of the industrial reserve and generally hostile not to capitalism but to socialism, their agency cannot be seen as positive.[69]

---

66 Basso (2021: 9).
67 See Bradley and Norhona (2022: 43, 44), whose misunderstanding both of surplus labour and of the free/unfree distinction fares poorly when compared with the more rigorous analysis by Pradella and Cillo (2021) of these same issues.
68 Mezzadra (2011a).
69 The difficulty faced by trade unions in uniting local and migrant labour to form a common front against capital was underlined in the course of a 14 May 2014 interview on the

Perhaps the oddest claim made by Basso is that, because the majority of immigrants to the EU have now found employment in the active army, they cannot any longer be considered to form part of the industrial reserve. Much rather, the case he makes illustrates precisely the opposite: the efficacy of the economic role surplus labour performs for capital. The fact that immigrants are now located within the active army demonstrates as clearly as need be the main purpose of the industrial reserve at work. They have displaced locals who either were previously employed in the active army, or else who hoped to enter the latter.

Equally problematic is his argument that migrants are in the vanguard of the struggle conducted by the working class, a view that contrasts with that of Marxists who have long maintained the opposite: that more often than not migrants are deployed by capital in 'from above' struggle *against* working class mobilization.[70] Thus the claim made by Basso, that both Marx and Lenin hailed emigrants to the United States as uniformly engaged in the class struggle alongside fellow workers overlooks the long history, in the United States of northern worker hostility not just to yet more immigration from Europe but also to slave emancipation, both on the same grounds: each would add considerable entrants to the labour market, thereby intensifying job competition to disadvantage of the existing workforce and to the advantage of capital.

Notwithstanding the claim that immigrants are in the forefront of anti-capitalist struggles in receiving countries, therefore, not mentioned by Basso (and others) are the less positive aspects of the open-door policy. These extend from the advantages gained by employers from continued access to the industrial reserve, including as a source of surplus labour used in strikebreaking and restructuring, to evidence that currently migrants – particularly from erstwhile socialist countries – do not want to change the system itself, only to get a better deal within capitalism as it is.[71]

BBC Radio 4. When asked why employers were recruiting workers like him in preference to locals, a recently-arrived migrant replied that he was prepared to do the same job for less pay. Asked, further, what he thought the impact of this would be on those currently in employment, he disavowed any sympathy for – let alone solidarity with – them, saying they would have to learn to live with such competition for jobs, and lower their wage expectations in keeping with the changed circumstances.

70 On migrants as in the vanguard of the working class, see Basso (2021: 7ff.).

71 The process of capitalist restructuring, involving the displacement of well-paid labour with cheaper foreign equivalents, was illustrated somewhat dramatically during March 2022, when in defiance of legislative procedure P&O ferries went ahead and in a single day sacked 800 British crew, immediately replacing them with stand-by low-paid substitutes recruited from Eastern Europe, an act justified by the CEO in terms of the need to remain competitive with rival companies. Significantly, this incident occurred just after

What Marx Really Said

Given his view about migrants being unproblematically in the vanguard of the 'from below' class struggle against capital, it is in a sense unsurprising that Basso is opposed to regulation of immigration and a supporter of an open-door approach. Attributing the latter to Marx, however, is incorrect. What Marx supported was co-operation between workers and working class organizations located in different national contexts, which is not the same view attributed to him by Basso: namely, support for open-door migration. In fact, Marx opposed this, as is clear from his views about the political and economic link between England and Ireland.

An argument frequently invoked by current supporters of open-door migration takes the form of internationalism which, it is claimed, not only licenses untrammelled worker mobility across borders, but is also a process that Marx himself endorsed.[72] Yet the interpretation of internationalism that Marx held was very different, as he made clear in a 1870 letter to Siegfried Meyer and August Vogt.[73] There he argued that the threat an increasing reserve army posed not just to hard-won wage levels and employment conditions but also to the protection of these gains – by means of solidarity among and capacity of an existing workforce to organize – was such that serious consideration was given by him to opposing further immigration. In order to stem competition from the industrial reserve army, and the way it permitted capital to divide-and-rule its workforce, therefore, a century and a half ago Marx advocated severing the link with Ireland precisely in order to prevent migrants from competing with and undercutting English workers.[74] He referred to the latter process as being

---

the pandemic and Brexit was marked by a contraction in the industrial reserve, leading in turn to labour shortages in the transport industry, as a result of which existing workers were able to negotiate higher wages and better employment conditions. What this episode underlines is that as soon as the bargaining power of labour increases, so capital resorts to the industrial reserve in order to lower costs and maintain profitability.

72 Those who conflate Marxist internationalism with support of open-door migration include Bradley and Noronha (2022: 69, 176 note 19), who argue – wrongly – that 'border abolition and anti-capitalism are one and the same, and both must be global and internationalist'.

73 See Marx and Engels (1934: 289–90).

74 On this issue, Marx (1934: 288, original emphasis) stated unequivocally that '[a]fter occupying myself with the Irish question for many years I have come to the conclusion that the decisive blow against the English ruling classes ... cannot be delivered *in England but only in Ireland*'.

'the secret of the impotence of the English working class, despite their organization ... a secret by which the capitalist class maintains its power'.[75]

Marx insisted that working class emancipation in England depended ultimately on Ireland following its own path of capitalist development, and to this end international solidarity would take the form of support from English workers for Irish equivalents in their struggle for economic and political independence, as distinct from migrating to where this had already occurred.[76] For Marxism, addressing the presence of industrial reserve army, together with related issues of its unregulated expansion, who benefits from this and why, as a prelude to its elimination, combine to form a crucial first step in any challenge to the accumulation process. It is only after this step that a government representing all workers (of whatever ethnicity and gender) can proceed to implement regulation of wages and conditions. Unlike Marxists, however, Basso accepts the latter objective, but rejects its being conditional on the realization of the former step.

Just how far from the concerns of Marxists the attempt to redefine the industrial reserve in positive terms is clear from the claim by Bradley and Noronha that '[a]nyone genuinely concerned about labour rights needs to understand that ... border controls only strengthen the hands of bosses'.[77] Aware of leftist opposition to open-door immigration, on the grounds that it benefits capital, Bradley and Noronha take issue with such views, arguing that '[i]t is worth restating some fundamentals of left politics'.[78] Although rightly pointing out that wage levels are determined by struggle, not immigration, and that what is needed is collective resistance by all workers, including migrants,

---

75 About the impact on class consciousness of this migration pattern, Marx (1934: 289–90, original emphasis) noted: 'Owing to the constantly increasing concentration of farming, Ireland supplies its own surplus to the English labour market and thus forces down wages and lowers the moral and material position of the English working class. And most important of all: every industrial and commercial centre in England now possesses a working-class population *divided* into two *hostile* camps, English proletarians and Irish proletarians. The ordinary English worker hates the Irish worker as a competitor who lowers his standard of life ... [t]he Irishman pays him back with interest in his own coin. He regards the English worker as both sharing in the guilt for the English domination in Ireland and at the same time serving as its stupid tool. This antagonism is artificially kept alive and intensified by the press, the pulpit, the comic papers, in short by all the means at the disposal of the ruling classes'.

76 According to Marx (1934: 290), therefore, '[t]he special task ... is to awaken a consciousness in the English workers that for them the *national emancipation of Ireland* is no question of abstract justice or human sympathy but the first condition of *their own emancipation*'.

77 Bradley and Noronha (2022; 57).

78 Bradley and Noronha (2022: 58).

their conclusion – that '[i]mmigration controls only weaken that capacity' – is incorrect.

The inference that borders undermine collective agency, and that abolishing such regulation empowers anti-capitalist organization, is no different from the view by Basso that those in the industrial reserve form the vanguard of working class struggle. Acknowledging that some unions oppose immigration, much like Basso, therefore, Bradley and Noronha then argue that 'trade unions are sites of struggle and we should be fighting for and within them'.[79] This contradiction stems from a failure to understand the reason for trade union hostility to surplus labour: the combined issues of more acute labour market competition, undercutting, expanding what is now a global reserve army of labour, all of which undermine the livelihoods of those in work, whose pay and conditions are themselves the achievements of long-standing class struggle with capitalists.

Travelling the Same Road?

Seen from the right of the political spectrum, the question of borders has risen up the agenda in metropolitan capitalist nations, making inroads into conservative government policies. Not the least of the many ironies is that, instead of slowing down migration, development in Third World countries has increased this. So long as *laissez-faire* remains dominant, and is accompanied by economic growth premised on open door policy, both skilled and unskilled labour-power in less developed areas will continue to migrate in search of higher wages, better-paid jobs, and improved working conditions available in metropolitan capitalist nations. This in turn has exposed the contradictions at the heart of conservative politics, both in the UK and in the wider system of advanced capitalism.[80] The endeavour in 2015 by British conservativism to regulate immigration in order to win back electoral support lost to UKIP, and the hostility expressed by employers to this, confirms that what capitalists want, now as in the past, is deregulated/unregulated access to cheap labour provided

---

79 Bradley and Noronha (2022: 59–60, 63).
80 This contradiction – '[p]eople are arguing against immigration but it's the only thing that's increased the potential growth of our economy' – was acknowledged late in 2022 by the Confederation of British Industry when emphasizing the continuing importance to the accumulation process in the UK of the industrial reserve, and calling for closer ties with the EU in order to allow further migration. See 'Business poised to anger Brexiteers by urging Sunak to "do the deal" with the EU', *Financial Times* (London), 21st November 2022.

by the industrial reserve.[81] This the UK conservative government from 2022 onwards has undertaken to provide, by easing restrictions on legal immigration, since when the latter process has merely increased, both numerically and in political importance.[82]

An issue that never really went away, immigration-as-industrial-reserve moved up the UK political agenda once again in late 2023, when the divide between on the one hand a conservative party wishing to retain the electoral support of its Red Wall seats opposed to immigration, and on the other employers lobbying for an open-door approach, became more acute following the publication of the migration figure for the previous year.[83] It revealed not just that net immigration in 2022 had increased to a record 745,000 annually, but also that virtually all of this was due to a 119% rise in work visas issued by government.[84] Underlined thereby were three things: that it was conservatives in power who were responsible for promoting this kind of labour market competition; that the latter was a cost-cutting exercise by capital, an outcome of migrants being paid only 80% of wage levels received by equivalent workers already employed in doing the same job; and, consequently, that the increase in labour market competition was a result not so much of illegal boat

---

81 In 2015 the UK conservative administration proposed to curb immigration in the name of anti-slavery, attempting thereby to regain those of its electoral base lost to UKIP. However, two things happened next. First, the backlash from many business organizations and thinktanks (IEA, CBI, IoD, the British Chamber of Commerce) all of which complained of the adverse economic impact that would result from no longer having access to cheap migrant labour. And second, the hasty backtracking by government which assured them post-Brexit exemptions would in fact allow continued recruitment/employment of migrant workers. This episode underlines the contradictory aspects of capitalism: a disjuncture between a political objective (anti-immigration to attract working class voters) and an economic one (pro-immigration to ensure capitalist profitability).

82 For continuing disagreement within UK conservativism from 2022 onwards, between a desire to boost economic growth by easing immigration restrictions, but to stay in power also needing electoral support from workers who want to see immigration controls, see See 'Liz Truss to review visa schemes in bid to ease UK labour shortages', *Financial Times* (London), 25 September 2022; 'Liz Truss plans to loosen migration curbs', *Financial Times* (London), 5 October 2022; 'Immigration policy cannot fix the job market', *Financial Times* (London), 26 October 2022.

83 On the divisions within the UK governing party over the immigration question, see 'Conservatives are on the hook for immigration', *Financial Times* (London), 7th December, 2023; 'Sunak is in a hole over immigration policy', *Financial Times* (London), 8th December, 2023.

84 See 'Sunak under pressure as net immigration reaches record', *Financial Times* (London), 24th November, 2023.

crossings – as conservatives and media commentators invariably claimed – but rather of legal migration facilitated and encouraged by the state.[85]

With few exceptions, however, the focus of political debate in the UK at this conjuncture remained largely on small boat crossings, avoiding the main cause of increased labour market competition.[86] Hence the way the issue was framed became one simply about outsourcing illegal migrants and their asylum claims to Rwanda, a discourse that in effect left unaddressed the reason for the growth in numbers. That notwithstanding, the role of immigration was to meet employer demands for open-door access to the industrial reserve was acknowledged occasionally, embodied in observations such as 'there has been an official willingness to rely on overseas staff to fill gaps in industries facing shortages, notably health and social care, rather than tackle wage and working condition issues that might make the roles more appealing to British staff', and 'immigration accounted for almost all the 0.9% increase in the size of the workforce in the year to the second quarter of 2023 … [f]oreign-born workers have plugged the gap left by a shortage of domestic candidates and so helped ease supply shortages'.[87]

From the left of the political spectrum, the issue of borders can – and should – be seen differently.[88] For Basso, as indeed for all Marxists, what is

---

85 Unsurprisingly, corporations resorted to virtue-signalling as a cover for taking advantage of the industrial reserve, maintaining that hiring policy would privilege refugees, 'far too many [of whom] remain unemployed, despite our endemic skill shortages, their high levels of education, desire to earn a living and legal right to work'. See 'Amazon leads drive to hire refugees', *Financial Times* (London), 19th June, 2023.

86 Because the industrial reserve contributes not just to the incidence of unfree labour-power but also (and therefore) to the undercutting of existing wages and conditions, it was acknowledged that attempts to rein in immigration might indeed be of benefit to those already in work. See 'Soaring wage growth and low joblessness strengthen workers', *Financial Times* (London), 14th June, 2023; 'Post-Brexit shift in immigration may mean higher wages and more self-sufficient UK economy', *The Guardian* (London), 10th September, 2023; 'Advisors urge tighter foreign worker rules', *Financial Times* (London), 4th October, 2023; 'Overseas care staff face long hours and large debts', *Financial Times* (London), 28th November, 2023; 'Advisers fear graduate visas fuel low-pay migration', *Financial Times* (London), 14th December, 2023.

87 See 'Rishi Sunak faces Tory backlash as net migration reaches record high', *The Guardian* (London), 25th May, 2023; 'Red wall Tory MPs put pressure on Sunak over net migration', *The Guardian* (London), 3rd July, 2023; 'Nurses from poor "red list" nations flock to UK', *Financial Times* (London), 30th November, 2023.

88 As net migration has risen to unprecedentedly high levels, even the UK Labour Party has finally accepted the negative impact that the industrial reserve has on the existing workforce. See 'Starmer to call for end of "low pay and cheap labour"', *Financial Times* (London), 22nd November, 2022; 'Net migration rises to record 504,000', *Financial Times* (London), 25th November, 2022. Despite invoking Beveridge as the way forward, however,

required nowadays is political opposition to capital by means of a joint struggle by each component of the working class – local and migrant alike – wherever the two co-exist.[89] This is uncontroversial, and a kind of mobilization all socialists can – and must – support. What Basso together with Bradley and Noronha fail to understand, however, is that as long as an open-door policy exists, capitalists will always be able to undermine – if not defeat – any such joint struggle by continuously recruiting 'green' workers from the industrial reserve. Not just locals but recently arrived migrants will, as soon as they unite and organize in pursuit of improved pay and conditions, be faced with the prospect of replacement by yet newer surplus labour drawn from this very source.

Hence the undeniably negative political impact of the industrial reserve: by its very nature (regulating the labour market on behalf of capital) it places limits on the success of any joint struggle. As one cannot emphasize too often, therefore, each component of the workforce (local + migrant) can and should unite, certainly, but – as Marx argued with regard to nineteenth century migration from Ireland to England – this unity should take the form of pushing for economic development and attacking capitalism within each context, to be followed by local/migrant unity within such contexts once the accumulation process had been deprived of continuing access to the industrial reserve. Rather than an open border or a sealed one, the object for socialists ought to be a tighter process of regulation based on planning that is no longer in thrall to a policy of enhanced labour market competition.

Conclusion

Historically, the impact of surplus labour on working class organization and struggle against capital has been perceived negatively by those on the left. For Marx, Engels, and other Marxists (Kalecki, Sweezy, Dobb, Glyn), the industrial reserve constitutes an obstacle not just to working class solidarity but also and therefore to the possibility of a socialist transition. Boosted by migration,

---

the more recent focus of the shadow Labour minister for employment and social security is not on the deleterious impact of labour market competition on wages and conditions – as did Beveridge – but rather on the need for a more efficient/'rational' capitalism ('[t]he progressive approach to failing markets is to intervene intelligently and make them work'). See 'A failing UK labour market requires state intervention', *Financial Times* (London), 14th December, 2023.

89 Basso (2021: 17).

it means that existing workers can be played off against incomers, and vice-versa: turning from locals then to migrants ensures that they are always in competition, generating acute hostility between different components of the workforce. Underwritten by the presence of the industrial reserve, it is a pattern of segmentation that permits capitalists not just to lower the cost of labour-power but also and thereby to compete more effectively by maintaining or enhancing profitability. This is what supporters of open borders mean when they say immigration is good for the economy.

For its part, liberalism has been more ambivalent, combining recognition of the negative aspects of surplus labour with acceptance of its positive contribution to the accumulation process. As long as unemployment exists, argued Beveridge, so also will acute competition for jobs, a result of which will be hatred of foreigners, Jews, women, and any others who seek to enter the workforce. However, he was writing largely about the UK jobs market in the 1930s, and how unemployment might be addressed by the British state, not about the labour market as a global phenomenon, as it has now become. Furthermore, Beveridge saw the solution as involving more capitalist production, generating additional employment, and consequently lessening the intensity of market competition among workers, all in a context of a nationally restricted and benign accumulation process, none of which applies currently. What he, like other liberals now, feared most was a 'from below' challenge to the capitalist system *per se*, and viewed the welfare state (and its provision) as a means of avoiding what he saw as a negative outcome (a questioning of capitalism and its replacement by socialism).

Unsurprisingly, one important effect of globalization has been to place borders, who crosses them and why, at the centre of debates about 'from below' empowerment. A consequence of this is that non-class identity is now inserted (or reinserted) into the question of labour market competition. The result is that all forms of immigration now tend to be recast ideologically by NGOs, Church organizations, postmodern academics and liberal journalism largely in non-economic terms, and consequently viewed not as a labour market issue – which is how Marxists and capitalists interpret it – but simply as a humanitarian one. A variant maintains that the industrial reserve is the locus of anti-capitalist struggle. According to these approaches, shorn of its negative attributes, surplus labour is instead redefined as positive.

For a number of reasons, such views are problematic. To begin with, migrants from ex-socialist countries would not be keen on socialist political transition, nor will they see a need to organize as long as wages/conditions in the receiving context are higher/better than in the sending one. Where they are unemployed, they can be used by employers – as P&O and other examples indicate – to undermine pay/conditions of those in work. Even in instances where migrants unite with locals in pursuit of better wages and improved conditions, as long as borders remain unregulated/open capital will always be able to replace those engaged in 'from below' struggle with 'green' labour. In short, the difficulty is the perpetual nature of labour market competition licensed by open-door policy, which in turn permits access by capital to the industrial reserve, generating in turn the rise and consolidation of resistance based not on socialism but rather on populism, nationalism, and 'nativism'.

What is missed by Basso, Mezzadra, Bradley and Norhona, and others, therefore, is that ideas about opposing immigration so as to block the industrial reserve were earlier advanced by the left, and – since the left no longer appears interested in them – have now been taken over by the right. This history is overlooked by those who simply equate criticism of open door migration with fascism and racism. Forgotten by many on the left is not just the way the industrial reserve has featured in the history of capitalism, and why socialists have viewed surplus labour negatively, as a gift to the accumulation process, but also and therefore the way it might feature in a socialist future. A crucial objective faced by an incoming socialist government is how to counter the anarchy of the market, the baleful legacy bequeathed by the capitalist system.

Because of this, socialism and socialists have always given priority to the regulation of the market – as much for labour-power as for other commodities – in order to facilitate central planning by the state. Economic planning advocated by socialists requires – indeed, depends on – strong control exercised over the element of freedom associated with the market (accurately embodied in the term *laissez-faire*), and workforce allocation by the state would of course be – inescapably – an important aspect of this. Hence the idea that cross-border free movement is in a very general sense politically emancipatory/progressive, and consequently would automatically flourish under a socialist government, is quite simply incorrect. As neither a transition to socialism nor, indeed, Marxism itself is still on the current agenda of development studies, having been displaced by varieties of postmodern theory, the issue of regulating – let alone reducing – the economic and political impact of a global industrial reserve does not at present feature among the policy objectives desired by a broad swathe of progressive opinion, including that of leftists. How the implications of this for the study of development play out is examined in the chapters which follow.

CHAPTER 3

Sociology and Development: a Warning from *The History Man*

> The bankruptcy of Christianity, pacificism, socialism and anarchism; it is always the intellectuals who have abdicated.
>
> An observation during 1925 by VICTOR SERGE on the negative role of the intellectual in the way all historical movements decline.[1]

∴

Introduction: Publishing, Hierarchy, Power

Academic publishing, and Marxist contributions to or exclusions from this, are possibly the most under-researched and undiscussed topics in the social sciences, which is ironic, given their centrality to the way in which intellectual discourse was conducted world-wide throughout the immediate the post-war era. The latter period, lasting for approximately the next three decades, saw the proliferation of academic journals, many of which were edited by Marxists, publishing much pathbreaking research and analysis on Marxism, in addition to debates generated and sustained as a result, is a matter of record. Perhaps the highpoint occurred during the 1960s, when the purpose of social science departments in new universities was twofold: to introduce, and then to undertake, the study and research into the dynamics of society, home and abroad.[2] It was an intellectual milieu in which development studies flourished. This was a period, moreover, that coincided with the shift in Marxism, from a practice at the level of the street, to its study at the level of the university. This process was itself accompanied by the entry of Marxists into academic posts, now required to provide the teaching about such political theory.

Accordingly, it is impossible to understand how academic publishing has operated in the past, and continues to do so still, without reference to this

---

1 Serge (2004: 34).
2 For a useful survey of the new universities during this era, together with their social science departmental background, see Pellew and Taylor (2021).

wider political and institutional background. The latter helps explain which, why, and when certain political approaches thrive or decline, a dynamic that is itself reflected in the shifting intellectual fashions taken up or discarded by academic journals, their publishers, and their readership. Not the least important aspect of this history is a consideration of the full extent and virulence of the political reaction to student protest at the new universities during the late 1960s, the purported role of sociology in generating this, and how such issues were depicted in popular culture, thereby constructing an image both of the new university sector, and of its emphasis on the social sciences and Marxism, for the wider society.

The reason for invoking the fictional narrative of *The History Man*, together with its failure wholly to anticipate that a fundamental change was about to take place in the dominant political discourse, is simple. Without addressing the manner in which a paradigm is – or is not – reproduced within the institutional structure of the university system, and how this in an important sense both influences and is influenced by academic publishing in all its forms (journals, books, edited volumes), it is difficult to situate the way in which a specifically Marxist publishing project might fare. This is especially true of the way editorial power is exercised via journals and books.

The presentation in this chapter is divided into two sections, the first of which examines how popular culture formed the negative image of sociology as taught at the 1960s new universities by portraying it as following Marxist fashion, thereby failing to anticipate the shift to the anti-Marxism of the cultural turn. The second section considers why such academic fashion is constructed and reproduced, together with its implications for the kinds of hegemonic trends encountered in social science publications.

I

The Bleak End of Things

Almost fifty years have passed since the publication of a campus novel *The History Man*, which is thought by some to have undermined – if not destroyed – the academic reputation and public image of sociology as learned in British university departments.[3] Written by Malcom Bradbury, who taught

---

3 On the negative impact *The History Man* is said to have had on the reputation of sociology and sociologists, see Ian Christie, 'Return of Sociology', *Prospect Magazine*, 20th January, 1999. For his part, Bradbury (2006: 144–45) rejects this view, arguing that he perceives the discipline itself as positive.

English Literature at the University of East Anglia, the novel went on to form the basis for a widely-acclaimed BBC television drama in 1981, solidifying in the public mind an association between what had hitherto been two distinct processes: teaching in the academic discipline of sociology; and a perception of the social sciences as licensing 'political indoctrination' by Marxism.[4] Tracing the negative effects of increased access to higher education, the narrative of *The History Man* portrayed the venal and self-serving pursuits of a radical leftist sociology lecturer – Howard Kirk – in a newly established university during the 1960s, in the course of which he advanced his own interests by exploiting in one way or another all those around him (students, colleagues, friends, family).[5] This is what happens, Bradbury appears to be saying, when anyone can become an undergraduate and a Marxist can end up teaching in a university.[6]

As in so many areas in academia, the historical focus on the 1960s new universities is reductive, and tends to lionize particular individuals, thereby

---

4   Whilst it is true that the view of sociology as a hotbed of Marxist theory came under attack from many other academics at this conjuncture – including Amis (1970: 157ff.), Cox and Dyson (1969-77), and Gould (1977) – it is nevertheless the case that the reach both of the novel and of the television version of *The History Man* was much wider in terms of audience and popular culture, as such being more responsible for the creation and reproduction of the negative public image, both of the discipline and of its institutional location.

5   The context and effect of this increased access to university education is outlined elsewhere by Bradbury (2006: 54–55): 'In 1960s Britain the Robbins Report was published, recommending a fresh expansion of higher education. Six new universities were built, the teaching of new subjects encouraged, and [student] grants even improved. At the time all this was seen as yet another fundamental revolution, probably a dangerous step ... In fact the new universities brought in much academic innovation, a variety of new subjects, syllabuses and teaching methods; but they still maintained the elite, selective, highly personal nature of British higher education. They also became smart and trendy places to be, competing with Oxbridge in the academic stakes. Around 1968 they also became rather radical places. The student revolutions that swept America and Europe found a special home in those pristine, architect-designed citadels'.

6   This view is in keeping with a broadly hostile attitude towards Marxism of sociology throughout its history as an academic discipline. In the late 1930s, for example, a sociology textbook (Lundberg, 1939: 73, 308, 409) dismissed Marxism as a 'system spun out of thin idealistic air', the 'dogmatic pronouncements' of which are unworthy even of study: an examination of its arguments 'is not contemplated', therefore, as 'the relative merits [of such a theory] is not here a relevant question'. Half a century later, not much has changed in this regard, the heterodox nature of Marxist theory where sociology is concerned still being the case. Conceding that '[i]t is important to try to approach the study of Marx's work in an unprejudiced way', Giddens (1989: 693, 701–2) maintains '[t]his is not easy', adding that '[i]t is best to see Marxism not as a type of approach within sociology, but as a body of writing existing alongside sociology'.

creating and then reproducing a cult of personality.[7] What happened at such institutions therefore becomes about what happened to one individual who is as a result cast in a heroic or anti-heroic role, not unlike the protagonist in *The History Man*. In the UK the new universities were set up in what now seems a comparatively benign post-war era, one informed by an expansion of higher education funded by state expenditure. The twofold object was on the one hand to provide the accumulation process with the requisite skilled labour-power it was thought to lack, and on the other with intellectuals whose 'problem-solving' knowledge would itself contribute to a more efficient capitalist production. Key to these objectives was the role of the social sciences generally, together with a focus on investigating the reasons for the economic backwardness of Third World nations.

Unmentioned by Bradbury, however, are two crucial and interrelated political issues. First, that students were not unthinkingly led by the nose, as in the way *The History Man* depicts the influence exercised by its main protagonist. Rather than following blithely what they were taught, therefore, the main target of the 1968 student movement was the conservative nature of the sociological theory then on offer. And second, the fact that many of those appointed to the academic posts in the new universities were at that conjuncture themselves products of the old ones, and brought with them the ideas, values, and politics of these ancient educational institutions.[8] One effect of this dissonance was a certain element of disdain towards student politics in general, and Marxism in particular, the inference being that it was presumptuous to attempt changing the educational system, let alone society.[9] It was

---

7   What students desired politically, together with their views about the way new universities, teaching, and society in general ought to change, quickly narrowed down to opinion expressed by 'representative' individuals. In the case of Essex University, for example, it was David Triesman who became synonymous with its 'student voice', featuring in most accounts of the 1968 protests at that university written by outsiders (Cockburn and Blackburn, 1969: 141–59; Widgery, 1976: 422; Fraser, 1988: 31, 61–2, 67, 110, 111, 114, 245–48). He subsequently followed a somewhat familiar rightwards political trajectory. Now a businessman and member of the House of Lords, Baron Triesman held political office in the 'New Labour' government of Tony Blair, and later resigned from the Labour Party radicalised under the Corbyn leadership.

8   A product of Oxford, one such academic taught sociology at Essex for a while before departing for a post at Cambridge, where his conservative political views surfaced in an episode recounted elsewhere (Brass, 2017b: 65–67).

9   The flavour of this contempt can be gauged from what was contained in the *Manifesto of Rationalism*, produced for a Revolutionary Festival that took place at Essex University in the 1969 Spring Term. Rumoured to have been composed by some members of the academic staff, its text included the following sentiments: 'My tone will be arrogant, but the arrogant were made so by the ignorant ... To every single student ... you are the vanguard of the vanguard of the vanguard. Raise on high YOUR banners emblazoned with the

in part against this kind of view, together with the far from politically radical sociology interpretations taught by such academics now lecturing at the new universities, that student protest was aimed.[10] Even when not employed in these new (= upstart) institutions, Oxbridge academics continued to influence what was – and what was not – taught, an exercise of power underlined by the Berlin/Deutscher episode.[11]

Despite not being a product of Oxbridge, Bradbury nevertheless subscribed to the kinds of traditional values such ancient university education represented.[12] Described by a close friend and fellow novelist as 'a liberal Tory' who not only enjoyed the 'pastoral life' but also distrusted 'modernity and the revolutionary desire for change', Bradbury espoused views consistent with the agrarian myth: supportive of the rural, the small-scale, and tradition, while opposed to the urban and modernity.[13] Unsurprisingly, therefore, these views

---

glorious legends of battles lost & prepare to lose again … Why do you bother? … March with Marx: the British Museum is with us; who can be against us? … Sociologist power, workers control (that is Control of the Workers by Commitees of properly qualified experts who will UNDERSTAND.) … students, and those who have been students, will be the first to be destroyed in "the revolution" [and] this is the main reason why they should seek to bring it about'.

10 On this point, see Cohn-Bendit & Cohn-Bendit (1968: 147, 168).
11 In what became one of the most notorious episodes of political discrimination, Isaiah Berlin, an Oxford academic, blocked the candidacy for a Chair at Sussex University of Isaac Deutscher, a revolutionary Marxist, and then proceeded to lie to everyone about his role in this decision.
12 Such values are prefigured in earlier writings by Bradbury. In a wide-ranging critique of 1950s affluence and modernity, he (Bradbury, 1962: 111, original emphasis) promotes and defends the idea of a return to the past, a view consistent with that expressed subsequently in *The History Man*: 'The person who lives in the country stands at least a chance of finding the traditional society's values at work, the old community sense … and even if he does not, he can at least learn to cultivate his own garden, in order to be ready when society does grind to a stop. It will be protested by the doctrinaire that such a person is living in the past, and sentimentalizing it. In the traditional society, however, there is no need to live in the past, since it is essentially the same as the present; it is only in a consumer society that one *can* sentimentalize the past, since it is different. Further, living in the past is better than living nowhere at all. In short, to choose, as a present-day man, to live in the past is to make a choice of thoroughly modern significance'.
13 On this see David Lodge, who in the 'Afterword' (Bradbury, 2006: 418) observes that '[i]n later life he enjoyed the occasional sojourn as a visiting fellow at Oxford [and he] always seemed very happy and at home in these settings – the smooth lawns, gravelled paths and ancient buildings soothed his spirit, and the ritual of hall and high table appealed to him', adding that 'Malcom was at heart a kind of liberal Tory … valuing tradition and pastoral life, tolerance and civility, distrusting modernity and the revolutionary desire for change'.

permeate *The History Man* narrative.¹⁴ In the book another lecturer, the political antithesis of the revolutionary leftist sociologist, is described by Bradbury as 'rural and bourgeois' who liked 'paddocks and stables', whilst prior to the expansion of the new university, 'this stretch of land was a peaceful, pastoral Eden, a place of fields and cows [where] the very first students, pleasant, likeable ... of quite another kind from the present generation'.¹⁵ The new university, by contrast, is depicted negatively, as antagonistic to the pastoral: 'all plate glass and high rise ... the campus is massive, one of those dominant modern environments ... that modern man creates', an urban context in which 'the peacocks have gone; the students are not bright originals in the old style' and '*Gemeinschaft* yielded to *Gesellschaft*; community was replaced by the fleeting, passing contacts of city life'.¹⁶

Not the least of the many ironies is that, shortly after the publication of *The History Man*, intellectual and political hegemony in academia passed swiftly from leftism to that of the anti-Marxist cultural turn.¹⁷ In part, this shift can

---

14 'The book had had a difficult gestation', accepts Bradbury (1987: 304), 'had come from an uneasy and pessimistic change in my own values ... as my initial excitement about the liberationist spirit of the 1960s moved toward a darkened unease'.

15 See Bradbury (1975: 39, 63–64), who elsewhere displays a similar kind of antipathy towards leftism which merely underlines the extent and significance of his political antagonism. Just as the target of his narrative in *The History Man* is a Marxist sociologist, so in his other fiction the object both of his satire and his censure are socialist governments of Eastern Europe (Bradbury, 1983, 1986). Hence the fictional Slaka (Bradbury, 1983: 49, 57) is 'the capital of a hardline country of the socialist bloc', a 'proletarian country' where nothing works properly. About its deficiencies all the usual tropes abound (Bradbury, 1983: 35, 38, 96): the awfulness of socialist regimentation ('evidently this is a culture where people are used to waiting'); the risible nature of attempts to account for economic inefficiencies; the uncomprehending yet rigid adherence of its population to state rules and instruction; a place where armed soldiers are everywhere, 'young men, with primal-looking unstated features'; and a worthless paper currency depicts 'muscular men wielding sledge-hammers and yet more muscular women tending vast machines'. In the words of a visiting academic (Bradbury, 1983: 37), '[t]here are colleagues of his at home who would regard [the people's republic of Slaka] as the model of the desirable future, the outcome to which a benevolent history points; there are others who would see it as the bleak end of things'. Unsurprisingly, it is to the latter category that Bradbury himself belongs.

16 Bradbury (1975: 64–65). Praising the novel and underlining its literary importance, Burgess (1984: 111) is surely wrong to commend its 'total objectivity'. Notwithstanding his view that '[i]t is a disturbing and accurate picture of campus life in the late sixties and early seventies', therefore, questionable is the conclusion that '[i]ts great aesthetic virtue ... is its total objectivity'.

17 Looking back on the political change that occurred shortly after the publication of *The History Man*, Bradbury (2006: 109) notes: 'The truly amazing thing is how in the last ten years, since Marxism collapsed and the Berlin Wall came down, the ideological divide by which a whole generation found intellectual seriousness has quite disappeared ... Now

be linked to the requirement for the skilled labour that students at the new universities would provide coincided with the beginning of deskilling, a result being that many of those who entered the job market with degrees thought to confer better employment prospects found that they were over-qualified. To some degree, it contributed to the subsequent dampening effect on radical leftist politics, a process that stemmed from an increased academic labour market competition and its attendant pressure to conform (= not rock the boat). It was this very same intake of students, radicalized politically in the late 1960s and early 1970s, that went on to provide the next generation of university lecturers, composed for the most part of leftists who, as soon as Marxism ceased to be fashionable, quickly and/or quietly abandoned it in favour of postmodernism.[18]

Who Is *The History Man* Now?

From the 1980s onwards, therefore, the dominant paradigm in the social sciences and the humanities more generally shifted dramatically; away from the materialist approach of Marxism, deemed inappropriate for an understanding of processes, issues, and populations outside Europe, and towards the 'new' populist postmodernism, the focus of which was on the empowering nature of identity politics. The latter approach was – and is – strongly antagonistic towards Marxist political economy, dismissed by postmodernists along with its conceptual apparatus of socialism/materialism/class as just one more kind of Eurocentric/Enlightenment 'foundationalism'.[19] Marxism was deemed to have nothing to say, either about the Third World, its pattern of economic development, or about issues in the metropolitan capitalist nations of the West.

Postmodern hostility expressed towards all things Marxist involves a twofold process: a denial of its historiography and conceptual apparatus is accompanied by an insistence on their replacement – epistemologically and politically – by a populist approach together with its privileging of peasant, ethnic, gender, and national 'otherness'. Marxism is declared irredeemably Eurocentric, tainted by a historical depriviliging of these same non-class identities that in the opinion

---

there is only one ideology [which] means none. We live in postmodern times. Ideas aren't beliefs but commodities. History's a theme park. Thought is irony. Liberal individualism turns out not to be a great humanist belief but pure capitalism, crass commerce, after all'.

18 On this point, see Brass (2017b: Chapter 18).
19 See, for example, the subaltern studies project associated with the work of Guha (1982-89).

of many postmodernists amounts to racism/sexism. It is this essentialist academic discourse, in effect recuperating and proclaiming as empowering all the categories and identities criticized hitherto by Marxist political economy, that it is argued here corresponds to the emergence of a very different History Man (and Woman), displaying all the negative characteristics attributed earlier by Bradbury to the main protagonist – a revolutionary leftist sociology lecturer – in his novel.

In one sense, what has happened in the years since the publication of *The History Man* in 1975 is the reverse of the process depicted by Bradbury. In his novel, a scheming and venal sociology lecturer who is revolutionary leftist carries all before him, and in a political dispute with the university authorities emerges triumphant, continuing in post and enjoying the political support of his students. Rather than the triumphalist consolidation of Marxism, those who adhered to the latter politics experienced something akin to a purge. The reality, therefore, was somewhat different. Leftists who sought or obtained university posts were faced with one of two options: either to water down or discard Marxism in order to gain a university job and then rise up academic hierarchy; or, sticking to principle by refusing to abandon Marxism, were denied access to or ejected from such employment.

Many went along with the first option, and in academic terms prospered accordingly, while those who followed the second path were denied the promotion merited by their scholarly achievements, remained on the margins of academia, or outside it altogether. The latter category included not just Isaac Deutscher but also others like George Rudé, E.H. Carr, Maurice Dobb, David Abraham, Jack Stauder, and E.P. Thompson. Unlike the protagonist of *The History Man* who, as a revolutionary leftist, successfully retained his academic post, therefore, those with a similar politics were in some instances prevented from getting permanent university jobs, from promotion once in them, or ejected from such employment as they already had.

Easy to forget is what can happen – and indeed has happened relatively recently – when such a political shift occurs in the wider society: for example, how in the 1990s many Marxists were dismissed from their academic jobs following the break-up of Yugoslavia ('A doctor of philosophy as a homeless person, a doctor of sociology as a horoscope writer, a journalist as a fisherman: these are the fates of Croatian Marxists after the "End of History"').[20] In

---

20   See 'The Lost Marxists: what happened to the academics made jobless by communism's collapse?', *New Statesman*, 23 November 2015, where – among many others – the following instances are recorded: 'Zvonko Šundov, a doctor of philosophy, got his last pay cheque 24 years ago ... The years he spent as probably the most educated homeless person in

such circumstances, political realignment signalled in journals and books can be seen as projecting the same kind of warning as the canary in the mineshaft.

In another sense, however, Bradbury rightly depicted two of the central dynamics at play during that era. The first of these was opportunism: what his main character, the radical leftist, pursues is personal advantage, disguised as revolutionary spirit.[21] The manner in which such opportunism is presented may indeed be exaggerated, but – unfortunately – it is all too true as a description of what happened when some leftists became tenured academics during that era. Hence the public stance that as university lecturers they would change the world for the better in some instances hid a desire merely to ascend the academic ladder, even to the apex of the university hierarchy.

The second dynamic rightly criticized by Bradbury was that of politics merely as academic fashion, a pervasive theme that surfaces throughout the novel.[22] Ironically, the 'Marxism' of the radical leftist sociologist as depicted in *The History Man* has little or nothing to do with Marxist theory and practice, being instead a case of 'do your own thing'. Rather than based on objective and rigorous theoretical analysis of issues in line with Marxist political economy, therefore, students are encouraged simply to follow a subjectivist/instinctual

---

Croatia have not broken him ... In 1991 [he] was fired from [his academic post and] has never returned to the classroom – because his job no longer exists. He taught Marxism ... Mira Ljubić Lorger, who has a doctorate in sociology, is another Marxist academic for whom the collapse of communism had dramatic personal consequences. Until 1990 she worked at the Social Sciences Research Centre in Split, a university institute which was, in the eyes of Croatia's new anti-communist government, a hotbed of Marxism'.

21 The issue of opportunism is also central to an earlier campus novel (McCarthy, 1953), set in the United States during the 1950s anti-communist investigations conducted by HUAC. Its main character pretends he is a communist so that liberals in the same university department are unable to sack him from his academic post without compromising their principles. Like the protagonist of *The History Man*, he not only behaves oppressively towards his students, but is described (McCarthy, 1953: 262) as an opportunist 'with a talent for self-dramatisation [and] one of those birds that are more communist than the communists in theory, but you'll never see them on the picket line'.

22 Hence the following kinds of accusation levelled at the main character (Bradbury, 1975: 16, 32, 40): 'You've lived off the flavours and fashions of the mind', 'you've substituted trends for morals and commitments', and 'I'm not wild about all this violent radical zeal that's about now ... [t]hey taste of a fashion'. From the outset he is described by Bradbury (1975: 3) as the epitome of the fashionable revolutionary leftist academic, 'a sociologist, a radical sociologist ... of whom you are likely to have heard, for he is much heard of [since] the university, having aspirations to relevance, has made much of sociology; and it would be hard to find anyone in the field with a greater sense of relevance than [him]. His course on Revolutions is a famous keystone' Towards the end of the novel (Bradbury, 1975: 228) he is labelled 'a radical's radical'.

path ('your own desires').²³ The picture that emerges is one of dilettantism: of 'Marxist'-academic-as-fake, of a revolutionary stance as essentially a fashion accessory, of learning as nothing more than 'a little Marx, a little Freud, and a little social history'.²⁴

That such an insubstantial approach is regarded by Bradbury as little short of dilettantism, and thus profoundly unintellectual, is clear. Despite coming to the new university 'with a reputation ahead of him ... for popularizing innovation', therefore, the novel's protagonist 'had not done a great deal of research on the book, and it was weak on fact and documentation'.²⁵ This element of dilettantism is itself reinforced by equating the attendance by the same revolutionary leftist at parties given by publishers in Bloomsbury, by socialists in Hampstead, and those given by new boutiques in the King's Road.²⁶ Underlined thereby is the central role of fashion, in ideas and politics no less than in clothes. Vogue in the case of the latter – cheaply made, all show, no substance, easily and quickly discarded – is a metaphor for the former being as much of a fad with similar kinds of characteristics.

II

The Power of Hierarchy

An effect of such a negative portrayal of the main character who, as well as being a Marxist and a sociologist, is also a venal and exploitative individual, is to condemn both Marxism and social science as innately hazardous – morally, politically, intellectually, ideologically – leading inevitably to the kind of behaviour exhibited by the protagonist of *The History Man*. It is a conflation that manages to produce or reinforce the impression that following fashion, opportunism, and lack of scruples are all outcomes simply of being a Marxist sociologist lecturer – and in the narrative are associated by Bradbury only with

---

23   When asked by a student for advice (Bradbury, 1975: 83–84), the leftist academic answers 'there's only one rule. Follow the line of your own desires'.
24   The theme of academic-as-fake is linked by Bradbury (1975: 69, 73) to the acceptance by the revolutionary leftist sociologist that all that was required intellectually was that 'you need to know a little Marx, a little Freud, and a little social history', a refrain that pops up a regular intervals, both throughout the narrative (Bradbury, 1975: 22, 26, 30, 119) and elsewhere (Bradbury, 2006: 143–44).
25   Bradbury (1975: 37).
26   See Bradbury (1975: 52), who writes later (Bradbury, 1987: 307) that 'I wanted to display in Howard Kirk the modern man of plots, something of a radical opportunist, living somewhere between the world of radical belief and that of fashion'.

Marxism and sociology – whereas palpably this is not so. Moreover, it delinks such negative attributes from another and more important cause: the competitive nature of the university as an institution operating within an increasingly neoliberal capitalist system, and the power exercised by those holding senior positions in the academic hierarchy.

Significantly, therefore, following academic fashion in the manner outlined by Bradbury in *The History Man* still persists, but now in a different way. Then – in the 1960s – it was about class, based on Marxism, whereas half a century later it is currently about privileging non-class identity. Class, together with Marxism, has not only long ceased to be fashionable, but become profoundly unfashionable among university departments and staff. Where academic publishing is concerned, fashion takes a specific form: jumping on the bandwagon when an argument, concept, or framework emerges. This is accompanied by an additional process: if the new interpretation happens to be formulated by Marxism, this is adapted by discarding its revolutionary agenda so that it fits in with bourgeois political ideology.

Following fashion, simply because it is fashion, usually entails accepting ideas, concepts or frameworks at their face value, invariably without interrogating their claims and origins. This practice can be seen at work in almost any social science journal, and especially those focussing on development studies. Not the least important aspect of following academic fashion is that it obviates the need for research into – and thus the questioning of – the claims advanced by a prevailing orthodoxy. Hence it permits the reproduction of an epistemological shortcut, to the effect that as a theory, an argument, a concept justifies the approach taken, there is little or nothing more needs to be said on this issue. There are many other reasons why this practice flourishes, not least the cult of the 'celebrity' academic and the deleterious impact on debate of academic seniority together with its kind of institutional power.[27]

Like so much else, it is impossible to examine why, in the case of journal publication, some kinds of approach are deemed acceptable while others are

---

27   In what was an accurate prognosis, the editorial in the first issue of a new leftist journal noted in 1987 that '[t]he New Realists of today have monopolised the media with the idea that there is no hope of a genuine socialist alternative. Only massive compromise and endless exercises in vote-catching are possible in the current political and economic situation, they claim … There is a danger of slipping into an alternative "star" system of Left celebrities, whose word becomes dogma; where debate turns into a monologue by the chosen few'. See '*Interlink* – a new magazine for the left', *Interlink*, No. 1, January/February 1987, pp. 2–3. This is exactly what happened, with some leftist journals vying for attention by diluting or abandoning core Marxist theory and/or practice, in the course of privileging the views of 'celebrity' contributors.

not, without reference to the power exercised within and beyond academic hierarchy.[28] Bluntly put, how often will an untenured junior lecturer go out on a limb by criticizing the approach of a senior and powerful member either of the same department, institution, discipline, or even of the journal to which an article has been submitted? To this question the only honest answer is almost never.[29] Evidence of this is hard to miss, particularly when one examines the bibliographies of journal articles. The latter frequently contain what might be termed reputation stroking by junior academics of their seniors, frequently misattributing to the latter the intellectual or theoretical advances in fact made by others.[30]

Another reason concerns a paradoxical effect of unexamined adherence to the dominant paradigm, which at a particular conjuncture enjoys the intellectual and political status of a 'given'.[31] Again ironically, this in turn can give rise to yet another kind fashionability, whereby a dissenting interpretation, once it gains ground imperceptibly, generates not just a plethora of mind changing revision on the part of others, but also spurious claims to have adhered to the dissenting view all along, or even to have formulated it in the first place. This academic phenomenon, too, can be seen at work in journal and book publication, in the form of attempted procrustean reformulation of past error so as to claim to always having been in step with what has now become a new orthodoxy. Instances of this kind of *volte face* are difficult to keep hidden from anyone carrying out even very basic research; that it does not often surface is, once again, down to the power exercised via academic hierarchy, where pointing

---

28 In what amounts to an academic equivalent of insider trading, some journals require submissions to focus exclusively on the arguments contained in publications by editorial board members, narrowing considerably the range and tenor of debate about the issue concerned. Other journals insist that book reviews say nice things about the tome in question, an instruction that undermines the very idea of criticism (on which see Chapter 10 in this volume).

29 The impact of power exercised in this fashion is structurally ubiquitous. In the world of film, therefore, Mamet (2023: 75, note, original emphasis) observes: 'Hollywood is hierarchical and savage. Those dealing with their inferiors may express an opinion, and the opinion may be disputed. But they are simultaneously proclaiming a *position*, criticism of which by an inferior is cause for dismissal'.

30 This is an issue that rarely surfaces in published form, an exception being Taylor *et al.* (2004) where it is referenced, albeit humorously (= a non-serious issue).

31 An illustration is the primacy allocated by the 'new' populist postmodernism to non-class identities (ethnicity, nationality) as progressive forms of empowerment, and as such the suitable – and indeed only – basis for political mobilization. To dissent from this approach, by pointing out its affinity with the ideology of the far right in the 1930s Europe, risks intellectual if not academic disbarment, and certainly enormous difficulties in securing the publication of such a view.

out in a journal article or a book review the inconsistencies/contradictions in arguments/claims made over time by a senior and institutionally powerful academic carries obvious career implications.

No One Is Listening?

Of related importance is a familiar trope within academic circles: that no publishing activity generates so much hostility as reviewing a book critically.[32] Grievances fuelled by less than effusive comments are legendary, and their effects can be long lasting. Inescapably, therefore, this applies with particular force to the kind of reviews that are not the sort which appear as quotations on the back cover of any subsequent editions of the tome in question.[33] The significance of this observation is that the impact on its writer of a damaging review is easy to underestimate, since where the offended author is a senior academic, the hostility is subject to a multiplier effect: in addition to the wrath of the author him/herself, therefore, all his/her friends, colleagues and/or clients are expected to show a similar level of upset.[34]

About this it is possible to draw upon personal experience. The recent publication by me of a review article pointing out the errors and misinterpretations

---

[32] That the impact of a critical book review can be considerable is recognized these days in the way some journals police this sort of context. Thus one prominent development studies publication requires that book reviewers include positive assessments, while another nominally Marxist one subjects book reviews to the same editorial refereeing procedure as a full article.

[33] Based on an erroneous chronology, some years ago a colleague accused me of leaving a trail of academic reputations unfairly wrecked due to my reviews and critiques, an inaccurate charge in that it overlooks the trajectory involved. With a few exceptions, such criticism has itself been a *response* to a previous and equally critical analysis of my arguments, and thus not a unilateral and unprovoked attack initiated by me. Defending one's views in this manner has always been central to debate over different interpretations about development theory and political economy, and cannot therefore be castigated plausibly as in some sense inappropriate.

[34] As has been outlined elsewhere (Brass, 2017b: 62–67), it was a critical review by Isaac Deutscher of a book written by Isaiah Berlin that resulted in the latter blocking the university employment prospects of the former. This extended from the candidacy of Deutscher for an academic post, to his participation in conferences, which confirms the operation of a process long known about but not often publicly acknowledged: namely, that one of the main drivers of social science discourse, and a major contributor to its political conservatism, is the power exercised by senior academics via their networks, pre-empting challenges to the status quo by subordinates for whom a display of too much heterodoxy might threaten career or promotion chances.

contained in an edited volume purporting to address what it termed critical agrarian studies, showing in particular how it had replaced a hitherto dominant Marxist paradigm with a populist one, elicited a symptomatic yet instructive response.[35] The reaction was somewhat predictable, taking the form of a communication from one of those whose contribution had been found wanting, to the effect that where my criticisms were concerned nobody in the development studies community was listening. The inference was clear: not that the criticisms themselves were wrong (they weren't) but simply – and egregiously – that they should not have been made.

Amongst other things, this kind of reaction demonstrates the inaccuracy of the self-serving myth that academia encourages – and, indeed, is based upon – critical endeavour in pursuit of knowledge. Apart from being incorrect, it reveals what is a common response to the publication of Marxist criticism: rather than engage with this, it is regarded as non-existent. In short, an erasure from the debate akin to an academic version of *damnatio memoriae*. It is hardly necessary to point out that an objection of this sort to a forensic theoretical and methodological engagement with a position in effect amounts to the forbidding of politics (= no criticism allowed). Contrary to what was claimed by this particular contributor, however, evidence suggests that in some cases those writing about development were – and are – indeed listening.[36]

---

[35] The edited volume in question is by Akram-Lodhi, Dietz, Engels, and McKay (2021a), reviewed critically by Brass (2023a). An updated version of the latter is included in this volume, as Chapter 4.

[36] In this connection, one notes merely that some analyses criticized by me have reappeared subsequently in a 'readjusted' version, in effect taking account of the comments made. For example, the shift by Jairus Banaji from an initial denial of feudalism and the categorization of unfree labour as 'so-called' (= non-existent) to the acceptance by him both of this mode and of unfree production relations. An analogous change by Jan Breman entailed the reclassification by him of unfree labour-power: from a production relation deemed incompatible with capitalism to its preferred form. Declaring Marxism wrong for not recognizing the acceptability to capital of unfree labour, and having been criticized for this, Genevieve LeBaron followed the same procedure, adopting the viewpoint of 'a more faithful Marxist tradition' – the very same view she had dismissed earlier. By contrast, Achin Vanaik eliminated his endorsements of Sumit Sarkar, also after the latter had been strongly criticized, whereas Utsa Patnaik – having initially declared the industrial reserve army an irrelevant concept when applied to India, and equally been criticized for this – did the opposite, and has now made the same concept the centrepiece of her case about the causes of poverty in India. Details about all these unacknowledged shifts, extending from critiques of the original argument to the appearance of the 'readjusted' version, are set out elsewhere (Brass, 2018a: Chapters 4 and 7; Brass, 2021b: Chapters 3, 5, and 6; Brass, 2022b: Chapters 3 and 4).

This kind of response has perhaps become more pervasive, given the current need to establish an individual space within what is an increasingly competitive academic market. Recognition that is conferred by being acknowledged as either having established an new approach within an existing discipline, developing a ground-breaking and original interpretation about an important issue, or alternatively reinterpreting the meaning of what has long been accepted as fact, has – as Hobsbawm once remarked – 'considerable compensations'.[37] Periodic disputes, by no means confined to the social sciences and humanities, about the origin of a particular idea or interpretation underline the importance of this issue, albeit one that – although tacitly conceded within academia – rarely surfaces in the public domain.[38] Should anyone be foolhardy enough to point out the error of such narratives – that claims about views held are incorrect, that intellectual discoveries are misattributed, and that the history/theory of development as presented is other than described – the reaction is akin to *lèse-majesté*.

Conclusion

Over a whole range of issues – among them the persistence and causes of economic crisis, the the pattern of changes in the capitalist labour regime, the deleterious political and ideological impact of an industrial reserve army that is global, the continuing importance of class, the political dangers of empowering non-class identities – the theoretical approach of Marxism has been proved right, time and again. Where academic publishing is concerned, this has generated contradictory responses. On the one hand, therefore, Marxism has been dismissed as outdated and irrelevant, as such having nothing to contribute either politically or economically to the analysis of present-day capitalism and the kinds of problems generated as a result. On the other, however, when a Marxist approach is shown to be right, the response of those who earlier cast doubt on its efficacy has in some instances been surreptitiously to adopt its argument and findings, without acknowledging this U-turn, hoping that this volte face would escape notice, or if it was would not dare to point this out. It

---

37 For this view expressed by Hobsbawm, see Evans (2019: 482).
38 How important this kind of recognition can be is clear from the fact that the dispute in the 1980s between American and French researchers over who first discovered the virus that caused AIDS was resolved only by intervention at Presidential and Prime Ministerial level (Reagan, Chirac).

is this latter option which corresponds to the kind of academic bandwagon-jumping satirized so effectively in *The History Man*.

Part of the difficulty in such cases stems from an earlier cause: during the 1960s the entry into academic posts of Marxists, and consequently Marxism as a topic of study. Mimicking the logic of capitalism, the process of competition/recognition within the university licensed what quickly became a plethora of reinterpretation. The latter entailed adding to what passed for Marxism concepts and theory that were non- or even anti-Marxist, leading inevitably to its dilution and depoliticization. Rather than the disempowerment of class, and its political resolution in the form of a transition to a revolutionary socialism, therefore, the desirable objective quickly shifted to the empowerment or re-empowerment of non-class identities, to be achieved without necessarily transcending the capitalist system itself.

All this poses difficulties for those who remain Marxists, in that they are tasked with interrogating not just the claims advanced by academic orthodoxy, but also by the holders of such views, an experience that understandably generates two sorts of discomfort. That felt by many senior academics when an attempt is made to examine how their political views have changed over time, the same being true of attempts to question the political credentials of purportedly leftist publications. Hence the institutionally disruptive practice and thus the unpopularity of the critical approach to existing theory undertaken by revolutionary Marxists. When the latter – unlike the character portrayed in *The History Man* – insist on interrogating claims advanced by exponents of anti-Marxist and/or bourgeois social science theory, they initially attract what is unmistakeably a torrent of opprobrium. In cases where such Marxist criticism is irrefutable, and can be seen as such, it is quickly and erroneously declared to be either commonplace or unconnected with the wider approach of Marxist theory. In the latter instance, the criticism is in effect deradicalized.

As with every other area subordinated to the rule of capital, academia is not immune to the power it exerts. Perhaps the most subverting influence is competition, a dynamic that drives both opportunism and fashion. As *The History Man* narrative makes clear, at stake is not just employment prospects but also research funding, book publication, and peer group recognition. Hence the need to keep up with, and certainly not depart from, the prevailing orthodoxy fuels a development that mimics the end result of the accumulation process: the establishment and reproduction within the domain of academic discourse in the social sciences of a trend akin to intellectual monopoly.

Unsurprisingly, therefore, a consequence is that a specific interpretation becomes hegemonic: as such it ceases to be interrogated, any attempt to do so inviting obloquy ('no one is listening'). This, it is argued here, has largely been the fate of Marxism during the epoch of the 'cultural turn', when the latter orthodoxy in the shape of the 'new' populist postmodernism assumed and exercised dominance over most forms of academic activity, not least employment, research, and publication. As will be seen in the following chapter, it is a process that has had a deleterious impact on the way development is interpreted by the social sciences.

CHAPTER 4

Critical Agrarian Studies and Development: a Populist Land Grab

> Silence is a species of artificial speech; it negates, it affirms.
>
> An observation by JOHANN GEORG RITTER VON ZIMMERMAN, about what silence cannot hide.[1]

∴

Introduction: the Sleep of Forgetfulness

If there is one issue that exhibits a symptomatic challenge to, if not a rejection of, the very idea of development, negating not just Marxism but also non-Marxist approaches, it is in the study of 'the rural' generally and of peasant economy and society in particular. Although like other areas of social science discourse the interpretation of agrarian change has never remained static, emphasis in analytical frameworks has shifted, both with prevailing ideological currents, not just in academic circles but also in the wider global context, and yet against what political economy teaches. Contrary to the latter approach, therefore, much contemporary discourse in development studies essentializes petty commodity production in an ahistorical manner, regarding rural smallholding merely as the economic pursuit of subsistence cultivation by an unchanging and undifferentiated peasantry.

From the turn of the millennium, the kinds of difficulties facing journal editors over the changes occurring in social scientific discourse and paradigms have multiplied. Although there has always been debate, and with it disputes about what a journal should publish, it could be argued that currently this issue is polarized more so than ever before. At the centre of this issue is what exponents of the 'cultural turn' present as innovative, state-of-the-art theory, claims about which have emerged and achieved intellectual prominence

---

1   Zimmerman (1800: 99).

over the past four decades.² The result is that within the social sciences – and indeed the humanities generally – Marxism has been marginalized, while a variety of concepts and approaches linked epistemologically to the 'new' populist postmodernism (subaltern studies, everyday-forms-of-resistance, global labour history, multitudes, empire, new social movements, ecofeminism, post-colonialism, post-Marxism, post-capitalism, etc.) have, in terms of intellectual fashion, become hegemonic.³ The role played by academic publications in this displacement has not been negligible.

Underestimated, therefore, is both the fact and the outcome of what might be termed 'journal capture'.⁴ A case in point is what occurred at one social science journal in 2008, when the editorial of a publication with a hitherto Marxist orientation (not absolute, however, as submissions following a populist line appeared from time to time) was replaced by one favouring a populist

---

2 Notwithstanding unpersuasive claims about 'newness', quite why the 'cultural turn' is – or should be – endorsed by so many sociologists, geographers, and anthropologists is a matter for conjecture. In the case of development studies, for example, the impact of this approach has been negative, in that the 'cultural turn' set out to deprivilege socialism, materialism and class as illegitimate Enlightenment/Eurocentric forms of 'foundationalism' inapplicable to the Third World. Quotidian resistance by (undifferentiated) peasants in defence of indigenous culture and tradition is instead seen as a legitimate part of the struggle against capitalism, a result being that rural struggle is no longer about class but identity politics. This recuperation of an essentialist peasant culture/economy leaves intact the existing class structure, and reproduces the populist mobilizing discourse of the political right. It is, in short, a conservative form of anti-capitalism.

3 In many ways, the 'new' populist postmodernism adheres to a Tolstoyan concept of peasant society, as is evident from the following (Tolstoy, 1905: 273): 'The only books that are comprehensible to the people and according to their taste are not such are written for the people, but such as have their origins in the people, namely, fairy-tales, proverbs, collections of songs, legends.' In what is the mirror image of the Marxist view, Tolstoy seems to imply that the rural masses are unable to transcend their own ideology, being anchored permanently in longstanding discourses that never change and cannot be altered. The latter interpretation is little different from the postmodern view of culture, with the difference that postmodernism *celebrates* this inability-to-transcend as the 'authenticity' of grassroots popular culture. As has been argued elsewhere (Brass, 2000), the conservative nature of this celebration by the 'new' populist postmodernism of existing (= 'traditional') plebeian forms – invariably in the name of an empowering 'from below' process of subaltern 'resistance' – is that it simultaneously consecrates and declares immutable the class system which gives rise to such culture in the first place, very much as Tolstoy – an aristocratic landowner – did in Russia.

4 At times such journals resemble nothing so much as a collection of little kingdoms, each one of which becomes the domain of an individual warlord and his retainers, all of whom jealously guard access to their newly acquired property. For their part, publishers don't seem to mind, so long as the new regime brings in a larger audience, which it does by following fashion and discarding controversy.

agenda.⁵ Since 2009, however, not just Marxist critiques of populism but also their contributions to debate have been marginalized. With much the same kind of braggadocio as the mid-1970s dismissal by cliometric historiography of earlier interpretations which categorized slavery as oppressive/exploitative, current populist academics declare Marxism banished from debate about the role of peasants in development.⁶ This populist view is based on a combination of shortcomings: an absence of knowledge about what constitutes Marxism; an unquestioning adherence to populist ideology; and – like cliometric historiography – a failure to interrogate the methods and assumptions informing both its own epistemology and that of the sources cited.⁷

These deficiencies, together with their implications for the way the role of peasants in the development process is understood, are nowhere so much on display as in a recent volume which attempts to redefine the whole field. The latter is the avowed project undertaken by a large portmanteau volume examined in this chapter – the recently published *Handbook of Critical Agrarian Studies* – that, according to its editors and contributors, covers the entire subject area as it currently exists.⁸ As such, it confronts any reader with two

---

5 The publication in question was *The Journal of Peasant Studies*. Much populist discourse, historically and currently, is simply about the desirability of reproducing a category of empowered petty commodity producers, who may or may not own the land they cultivate, a crucial distinction that often remains unclarified. This obscures the presence of what are in many rural contexts antagonistic class positions. Petty commodity producers who employ labour can be – and often are – as harsh employers as a landlord, not least because they in turn are subject to the most acute forms of market competition (and thus have to keep their costs to a minimum). For their part, landless workers seek to improve both pay and conditions, and for this reason are frequently in conflict with employers per se. This distinction is obscured by conflating petty commodity producers and landless workers in the same populist category of 'the rural poor'.
6 See, for example, the erroneous claim (Levien, Watts, and Hairong: 2018: 853–54) that 'much of this debate – and in fact a good deal of the earlier peasant studies research of the 1960s and 1970s – has reached something of an impasse. In part the impasse arises because new empirical work addressing the complex contemporary patterns and conjunctures of global agrarian capitalism, and because new and generative theoretical reconstructions of Marxism itself, offer exciting new analytical horizons'. What are termed 'exciting new analytical horizons' frequently turn out to be nothing of the sort, recycling old populist tropes recuperated unknowingly by present-day adherents of the same theory.
7 Problematic methods were central to the forensic dismantling by Sutch (1975) of the revisionist case about the non-oppressive nature of plantation slavery in the antebellum south made by the cliometric historiography of Fogel and Engerman (1974). For a critique of methods used in support of similar kinds of claims made by agrarian populism concerning Russia and India, see Brass (2020).
8 Edited by Akram-Lodhi, Dietz, Engels, and McKay (2021a), the volume contains 700+ pages, consisting of 72 chapters written by 91 contributors.

obvious questions. First, what – and who – is not in it that should be, and what is in it that shouldn't be, or the nature of the inclusion/exclusion pattern structuring the way the subject is presented; and second, how accurate is the resulting picture it draws. In short, whose voice are we hearing, why, what is it saying, together with the resulting impact on the way the whole approach to the subject area in question is framed and discussed.[9]

For the most part, chapters in the *Handbook* are composed by those with an interest in ecology, the climate, and the environment, but with little knowledge either about the agrarian question and its relevant debates, or about Marxist political economy, its concepts and theory. Many of the chapters are compiled by recent contributors to one or both main journals concerned with peasant economy/society, which helps explain why the focus is largely on hymns of praise to agrarian populism and its current exponents.[10] It also helps explain why significant Marxist contributions to debates about the peasantry are missing, sidelined or misinterpreted, as dissenting voices remain unheard or marginalised. Explored here, therefore, is not just the misleading way in which aspects of the agrarian question, together with its related issues/concepts, are presented, but also why this is so.

Broadly speaking, the agrarian question, and its form of resolution, is central to any understanding of what happens to the peasantry in a developing society, and why. As such, it informs much historiography and political economy discourse concerning the presence or absence of economic growth, together with its political causes and effects. Marxism frames the agrarian question in terms of systemic progress towards socialism; other than establishing or

---

9   Over a century ago Lenin (1964c: 46–47) outlined both the role and the shortcomings of such volumes in the following manner: 'The whole task of a handbook of political economy is, of course, to give the student of the science the fundamental concepts of the different systems of social economy and of the basic features of each system; the whole task is one of placing in the hands of the student ... a reliable guide to the further study of the subject, so that, having understood that the most important problems of contemporary social life are intimately bound up with the problems of economic science, he may acquire an interest in this study. In ninety-nine cases out of a hundred this is precisely what is lacking in handbooks of political economy'.

10 As outlined elsewhere (Brass, 2000, 2014b), agrarian populist discourse is structured by a symptomatic combination of likes/dislikes. It approves of 'natural'/harmonious rural-based small-scale economic activity (peasant family farming, handicrafts) and culture (religious, ethnic, national, regional, village, family identities derived from Nature). By contrast, agrarian populism disapproves of urban-based large-scale economic activity (industrialization, finance capital, the city, manufacturing, collectivization, planning, massification) and its accompanying institutional/relational/systemic effects (class formation/struggle, revolution, socialism, bureaucracy, the state).

re-establishing subsistence-oriented family farms, however, populist theory about systemic transformation appears vague or purposeless. In an important sense, therefore, claims in the *Handbook* regarding both the logic and the political direction of agrarian studies (critical or not) stand or fall depending on the way the agrarian question – in both its classical and contemporary versions – is framed and resolved.

What follows is composed of two sections, the first of which critically examines the claims made in the *Handbook* regarding the newness both of its approach and of its break with the past. The second considers difficulties arising from the way the agrarian question is interpreted, together with the reasons for this.

I

In the Academic Salon

As is so often the case, problems emerge right from the start. The introductory chapter, by Akram-Lodhi, Dietz, Engels, and McKay traces the origins of what it terms critical agrarian studies (CAS) to the *Journal of Peasant Studies* (JPS) and the *Journal of Agrarian Change* (JOAC), the two journals most closely associated with research into, debates about, and the conceptualisation of 'the rural' in its broadest sense.[11] Part of critical development studies, CAS is described as critical 'in the sense that it seeks to challenge dominant frameworks and ideas in order to reveal and challenge power structures and thus open up possibilities for change'. Retaining aspects of Marxist political economy, including class, CAS 'newness' is attributed by Akram-Lodhi, Dietz, Engels, and McKay to breaking with earlier approaches characterizing peasant studies. Unlike the latter, therefore, the 'newness' of CAS takes the form of combining micro- and macro-level analysis, and paying attention to previously unaddressed 'sociocultural dimensions'. Described as 'a highly diverse and emerging interdisciplinary field', CAS is said by Akram-Lodhi, Dietz, Engels, and McKay to embody a new 'pluralism'.

As to what CAS is against, the 'dominant paradigm' to be challenged is, we are told, 1950s modernisation theory which argued that economic development would depeasantise smallholders by converting them either into 'entrepreneurs' or wage labour. 'Scholars from [CAS]', observe Akram-Lodhi, Dietz, Engels, and McKay, 'do not accept this paradigm'. Accordingly, over the past

---

11   Akram-Lodhi, Dietz, Engels, and McKay (2021b: 1–7).

quarter of a century there has been a shift from peasant studies to CAS; along with 'classical analysis of the agrarian question', Marxist political economy has diminished in importance, as a result of being displaced by 'social science orthodoxies'. It was this 'unravelling' which led to the 'emergence of [CAS] as a field of study'. Unfortunately, the account in the Introduction about what CAS claims to be, and why it is different from what went before, is wrong in every respect.

Deprivileging Marxism

To the claim that what makes CAS novel is its critical approach, a departure from previous development theory, one is tempted to reply: when has the latter not been informed by a critical approach? Such hyperbole deployed merely to justify the specificity of CAS faces a number of difficulties. To begin with, it overlooks the fact that most – if not all – development theory hitherto has involved critiques of one sort or another. Furthermore, the combined micro- and macro-structural methodology plus 'socio-cultural dimensions' deemed absent earlier long predated CAS, as a glance at the subjects covered by the two JPS Indexes (volumes 1–31, spanning 1973–2004) would reveal. The assertion that CAS is opposed simply to 1950s modernisation theory hides the fact that the latter includes Marxism, the development approach of which – like non-Marxist variants – also embraced the idea of modernisation. Indeed, this very point is conceded in the Foreword, which similarly objects to a 'persistent modernisation narrative', noting, however, that CAS 'begins from a different premise [and] is critical of versions of modernization *and Marxian theory* which endlessly rehearse transition narratives' (emphasis added).[12] In other words, the Foreword gives the game away by revealing that Marxism forms part of the modernization framework to which CAS is opposed. For this reason, the insistence that CAS is sympathetic towards – and thus still incorporates – a Marxist framework is open to question. Much evidence, both in the *Handbook* and elsewhere, underlines the extent to which the relationship between CAS and Marxism is an antagonistic one.[13] And last, it fails to identify the real element

---

12   Li (2021: XXIII).
13   The vehemence with which Marxist approaches are dismissed elsewhere by CAS adherents (Hall, Edelman, Borras, Scoones, White and Wolford, 2015: 469–70) is the following: 'numerous lamentations about "false consciousness", "hegemony" or failure to recognize the collective's "true" … interests … The Marxists' bemoaning of "false consciousness" … Peasants' "failure" (for many Marxists) …to transcend "defensive" local struggles (for

of difference, a political one, informing the 'newness' of CAS: a deprivileging of Marxism combined with a reprivileging of agrarian populism.

This deprivileging/reprivileging is the key to the way debate on the agrarian question is framed by CAS. Marxists perceive class struggle as leading to depeasantisation, the formation of a proletariat, and a transcendence of capitalism that takes the form of a socialist transition; agrarian populists by contrast see the political task in hand as being repeasantisation, the establishment and protection of smallholding agriculture, licensing either a return to a precapitalist social order or to a 'nicer' sort of capitalism. Given the populist tone of the *Handbook*, it comes as no surprise that the chapter on peasants is by van der Ploeg, a strong advocate of just such an approach.[14] He commends the resistance theory of Scott ('weapons of the weak'), a central element of populist discourse, arguing that 'new representations' of rural smallholders include 'the capacity of the peasantry to articulate itself as a "class for itself"'.[15] This claim – much criticized by Marxists, including me – not only reproduces two populist tropes – equating peasants with a class, whose agency is consequently a form of class struggle – but then incorrectly situates them within a Marxist framework.[16]

The extent of misunderstanding what is – and what is not – Marxist theory can be gauged from the chapter on class by Berry, where it is stated that 'people are exploited by the political, economic and social world they live in as well as by members of a particular class', attributing to 'the world' an all-embracing capacity to exploit.[17] That the class differentiation of the peasantry, the model associated with the Marxist interpretation of Lenin, is 'integral' to CAS, as claimed by van der Ploeg, is similarly disputatious.[18] The same difficulties arise in the case of the debate about the way production relations change, and why. Unfree labour, which Marxism defines in opposition to its free equivalent, is instead relegated by Pattenden to a problematic bit-part category on an employment 'continuum', in the process failing to note – as does Harriss – the contradictory pronouncements by Breman on the connection between capitalism and debt bondage.[19] A better account of the same capitalism/

---

social movements theorists), became just one more indication of their atavistic politics and culture'. For opposition to Marxism encountered in the Handbook itself, see below.
14   van der Ploeg (2021: 109–119).
15   van der Ploeg (2021: 112, 113).
16   For critiques of Scott, see Brass (2000; 2017b: Chapter 12).
17   Berry (2021: 70).
18   van der Ploeg (2021: 109ff.).
19   See Pattenden (2021: 93) and Harriss (2021: 412, 413). On the many contradictions encountered in the case made by Breman over the years about the connection between unfree production relations and capitalist development, see Chapter 8 in this volume.

unfreedom relationship, however, is provided by Gerber, the only contributor who refers to the Marxist concept of deproletarianization.[20]

Although CAS likes to present itself as a radical alternative to and departure from orthodoxy, its populism is nevertheless an accurate reflection of the drift towards post-1980 mainstream academic conformity, firmly in step with conservative politics based on individual property and so-called 'popular capitalism'.[21] This populist agenda is itself represented in the chapter titles, indicating opposition towards the process of depeasantisation (land grabs). The latter, however, is based not on a materialist analysis to do with historically specific kinds of class formation/consciousness/struggle, aimed at systemic transcendence and establishing collective ownership of property, as in the case of Marxism. Instead, CAS populism is based simply on vague non-systemic/ahistorical moral/ethical concepts (the right to food, agrarian justice, human rights) deployed in support of individual private property ownership, peasant economy, and subsistence-oriented cultivation ('food regimes', 'food sovereignty', 'food security').

Current Marxist critiques of this approach, many published in the pre-2009 JPS, receive scant attention (see below): lacking, therefore, is a sustained consideration (and frequently even a mention) of negative appraisals featuring agrarian myth discourse, the 'new' populist postmodernism, and the subaltern studies project.[22] At most, there are brief and undefined references, both by Oliveira and McKay, and by Dietz, to 'subalternity' and 'subaltern actors', accompanied by uncritical endorsements of 'post-colonial' theory.[23] In what is a laudatory account of postmodernism, Leinius celebrates its overlap with CAS, declaring enthusiastically that postmodernism can 'only strengthen the critical impetus and work of critical agrarian studies'.[24]

---

20 Gerber (2021: 549, 550).
21 This politically conservative trajectory, together with the reasons for its emergence, is outlined in Brass (2018b). That CAS does not deviate substantially from what has long been an orthodox bourgeois mainstream development framework is evident from, for example, Ellis (2000) and de Janvry and Sadoulet (2011).
22 These negative appraisals include, among others, critiques by Nanda (2001; 2004), Petras (1990; 2002), and Brass (1997a; 2000; 2014b; 2020). Marxist critiques of populism are not the only absences, since missing from the *Handbook* are references to important political analyses both by historians of agrarian populism and by populists themselves (Sorokin and Zimmerman, 1939; Mitrany, 1951; Venturi, 1960; Wortman, 1967; Laclau, 2005b), whose views might usefully have been compared to – and contrasted with – those held by current exponents, not least by underlining the lack of CAS 'newness'.
23 See Oliveira and McKay (2021: 321) and Dietz (2021: 606).
24 Leinius (2021: 610–617).

II

'Marxist' Questions

Problems with interpretations of what is and what is not Marxist theory surface prominently in the *Handbook* chapter by Watts on the agrarian question.[25] Having designated his chosen texts (Marx, Engels, Kautsky, Lenin, Chayanov, Preobrashensky) as canonical, Watts then not only attempts to portray Kautsky as a prefigurative ecologist but also counterposes him to Lenin. According to him, therefore, Kautsky is an adherent of the same position as that held by the neopopulist Chayanov – that peasant economy is able to reproduce itself even where the agrarian sector is penetrated by capitalism – as against the view of Lenin that capitalism necessarily differentiates rural petty commodity producers along class lines.[26] Observing that for Kautsky 'the smallholder commanded centre stage' due to an 'ability to resist competition', Watts maintains that the Marxist theorist responsible for the 1899 *Agrarfrage* departed from the peasant differentiation/vanishing model, in effect negating the case made by Lenin at the same conjuncture. In the opinion of Watts, 'Kautsky's intervention proved to be brilliantly prescient and a sort of theoretical and political challenge to Marxist orthodoxy'.[27]

That the agrarian question formulated by Kautsky was in essence no different from that of Chayanov, whose interpretation it upheld against that of Lenin, is quite simply wrong. Overlooked is the reason given by Kautsky for 'peasant persistence': not, as Watts supposes, an 'ability to resist competition' – the populist claim about the efficiency of peasant economy – but due to an entirely different cause. Smallholdings were not expropriated since they were the source of cheap (and often unfree) labour-power required by large estates or agribusiness enterprises, drawn by the latter from the peasant household and/or resident kin. It was this, and not any innate economic viability, that determined the survival of smallholdings, and why such units are never wholly displaced by large commercial producers.

Furthermore, this was an explanation that met with strong approval from Lenin himself, who perceived no contradiction between his own view and that of Kautsky. In a review of *Die Agrarfrage* Lenin endorsed the analysis of Kautsky, noting that 'it would not even be advantageous for the big landowners to force out small proprietors completely [since] the latter provide them with hand! For this reason, the landowners and capitalists frequently pass laws that

---

25 Watts (2021: 53–67).
26 The main interpretations of the agrarian question, each of which appeared in the same year, are by Kautsky (1988) and Lenin (1964b).
27 Watts (2021: 55).

artificially maintain the small peasantry'.²⁸ *Pace* Watts, who attributes 'peasant persistence' to smallholder 'ability to resist competition', for Lenin – as for Kautsky – the opposite held true: 'Petty farming becomes stable when it ceases to compete with large-scale farming, when it is turned into a supplier of labour-power for the latter'.

Similar difficulties arise as to notable Marxist contributions to the agrarian question missing from the list of those Watts regards as forming the canon. No mention is made of Trotsky, despite the centrality of peasant agency to his theory of permanent revolution, an interpretation he formulated as an alternative to Stalin's two stages model of agrarian and systemic transition.²⁹ Because rural smallholders were seen by those hostile to socialism both as upholders of private property, and as bearers of national identity and traditional culture, Trotsky warned against the political objective of promoting yet more capitalism in the countryside, since it was already there.³⁰ His view was that struggle ought instead to be for a direct transition to socialism based on the dictatorship of the proletariat – not an intervening bourgeois democratic stage – on the grounds that, once they obtained land, rich and middle peasants would

---

28 See Lenin (1964b: 96–97), who expressed undisguised enthusiasm for the case made by Kautsky, calling its publication 'the most important event in present-day economic literature since the third volume of Capital', and concluding (Lenin, 1964b: 98): 'Applying the results of his theoretical analysis to questions of agrarian policy, Kautsky naturally opposes all attempts to support or "save" peasant economy'. Lenin went on to defend the agrarian question as interpreted by Kautsky against critics of the latter (Bulgakov, Struve).

29 See Trotsky (1962: 7–8), who points out that permanent revolution addresses 'the problem of the transition from the democratic revolution to the socialist', in view of the fact that 'a bourgeois counter-revolution [only serves] to preserve pseudo-democratic forms [whereas] the dictatorship of the proletariat puts socialist tasks on the order of the day'. Later on he elaborates (Trotsky, 1962: 154) that the 'endeavour to foist upon the Eastern countries the slogan of the democratic dictatorship of the proletariat and peasantry, finally and long ago exhausted by history, can have only a reactionary effect. Insofar as this slogan is counterposed to the slogan of the dictatorship of the proletariat, it contributes politically to the dissolution of the proletariat in the petty-bourgeois masses [the peasantry] and thus creates the most favourable conditions for the hegemony of the national bourgeoisie and consequently for the collapse of the democratic revolution'. Details of the reasons for interpreting the political role of petty commodity producers in this manner, together with the kind of political links between peasants and workers licensing a socialist transition, are set out elsewhere (Trotsky, 1962: 201ff.).

30 Unlike populists, who insisted that because capitalism was absent from the countryside, it could be avoided, and that village society was anyway based on socialist principles, Trotsky (1934: 419) commented: 'All Narodnik theories to the contrary notwithstanding, there was not in this [peasant programme] one grain of socialism. The most audacious of agrarian revolutions has never yet by itself overstepped the bounds of the bourgeois régime'.

oppose further socialisation of the means of production aimed at converting all private holdings – not just those confiscated from a landowning class – into state property.³¹

These problems are linked in turn to the populist sympathies displayed by Watts himself. Hence the manner in which he describes Marxist critiques, along the lines of 'purported weaknesses, romanticism or untenable natures of peasant populism', in which the revealing term 'purported' hints at disagreement.³² In arguing that 'the diversity and dynamism of agrarian questions ... seem as vital as ever [since] the agrarian question is very much alive and kicking in the 21st century', Watts appears to disagree with Bernstein.³³ Nevertheless, the reason for this is because he misreads Bernstein, maintaining the latter argues that as smallholders 'are in effect no longer peasants or petty commodity producers' they have to be seen as 'semi-proletarianised', the inference being that as a result the smallholder must be seen as a worker rather than a peasant. Watts appears to suggest that in depriviliging 'peasantness', Bernstein's interpretation is akin to that of Lenin. The perception of Bernstein, however, is exactly the opposite: 'The starting point must be to view peasants today as agrarian *petty commodity producers within capitalism*', adding that '[t]his is also the position of more sophisticated populists'.³⁴ Like Akram-Lodhi and Kay, Watts also refers to the work of Gibbon and Neocosmos without noticing that it is highly critical of Bernstein's agrarian populism.³⁵

'Marxist' Answers

If the CAS interpretation of the classical agrarian question (Lenin, Kautsky) is flawed, then its reconceptualization by contemporary exponents is no better. Many contributions to the *Handbook* tend to frame current discussion on the agrarian question simply in terms of arguments put forward by Byres and

---

31 Unsurprisingly, perhaps, the single *Handbook* reference (Lafrance, 41) to the work of Trotsky is unconnected with his views on peasants. Together with those of Lenin, the opinions expressed by Trotsky are the most damning political critiques by a Marxist of populist theory.
32 Watts (2021: 58).
33 Watts (2021: 61, 62).
34 This view of Bernstein as a self-confessed populist is cited in Brass (2007: 139 note 30), the source of the quote being Bernstein (1990: 72, original emphasis).
35 Akram-Lodhi and Kay (2021: 26), Watts (2021: 59). For the argument that Bernstein is an agrarian populist, see Gibbon and Neocosmos (1985).

Bernstein, and subsequently by Borras.[36] With few exceptions, their take on the issue is endorsed unquestioningly, despite the existence of critiques that fundamentally challenge such explanations. Not mentioned, therefore, are the political and theoretical problems that arise as a result. Unlike the teleology structuring the 'classical' agrarian question, which in the case of Marxist theory involved historical processes, political conditions and social forces that prefigure socialism, this kind of transition – the systemic transcendence of capitalism leading to socialism – has vanished from the agrarian question as conceived by Byres, Bernstein and Borras. Difficult to miss are the problems which accrue from this failure to question such interpretations.

An exponent of the semi-feudal thesis, Byres restricts the agrarian question to capitalist development within a given national context.[37] Unless labour-power employed there is free, capitalism is deemed by him to be absent or insufficiently developed, a view which discounts accumulation by international corporations using unfree workers to restructure the labour process. Since unfree labour is misinterpreted by the semi-feudal thesis as a 'pre-capitalist' relation, its presence signals wrongly that a transition is to be to yet more efficient capitalism, not socialism. Confining the agrarian question to national contexts, therefore, allows capitalism off the hook, banishes socialism from the political agenda, and permits the myth of a 'progressive' national bourgeoisie to flourish. This reformist – not to say conservative – approach of Byres is the problematic way in which the agrarian question is resolved in his semi-feudal analysis.

As in the case of Byres, the final stage of political transition has also vanished from the agrarian question as seen by Bernstein, who expels from the development agenda both a socialist project and its subject, the working class. In an attempt to re-invent the classic Marxist analyses of Lenin and Kautsky, Bernstein maintains implausibly that there are not one but two agrarian questions: that of capital (= modernization), and that of what he terms 'labour'.[38] His view is that the former has been completed, while the latter has not, and is now a question of 'classes of labour'. Rejecting the Marxist approach that

---

36 The CAS volume contributors restricting the agrarian question to what is said by Byres and Bernstein include Friedmann (2021: 15ff.), Akram-Lodhi and Kay (2021: 26), Jan and Harriss-White (2021: 171), Greco (2021: 251), Jha and Yeros (2021: 335ff.), and McKay and Veltmeyer (2021: 504ff.).

37 On this interpretation of the agrarian question, see Byres (1996). Despite not being a populist, Byres' take on the agrarian question is nevertheless flawed.

38 For these views, see Bernstein (1996/97, 2006) His concept 'labour' is never properly defined, oscillating between 'labour'-as-(undifferentiated)-peasants, 'labour'-as-workers, and 'labour'-as-any/everyone-who-works-manually-on-the-land.

differentiates the peasantry along class lines, Bernstein pronounces Lenin mistaken, and instead argues for the replacement of Marxist concepts such as 'proletariat'/'proletarianization' with his own term 'classes of labour', a category which includes undifferentiated peasants.³⁹

Eliminating modernization from the agrarian question in this manner, and substituting 'classes of labour' for proletarianization, suggests the possibility that what Bernstein terms 'labour' may possess interests/objectives that transcend history rather than being determined by it. This is what populism argues with regard to the peasantry, the characteristics of which are perceived as immanent and unchanging, regardless of the economic system in which smallholders are located. The suspicion remains, therefore, that lurking behind the term 'labour' purged of modernisation is actually peasant economy, and that the object of dividing the agrarian question in two is to retain within development discourse something akin to a Chayanovian model. Why these Marxist concepts and processes are discarded, and why smallholding is retained within his all-inclusive sociological category ('classes of labour'), is not difficult to explain: it is because Bernstein is himself an agrarian populist, not a Marxist.

These difficulties notwithstanding, contributions to the *Handbook* are generally supportive of such interpretations, for the most part unexamined by contributors. Among the 'major interventions' in the agrarian question debates of the 1970s/1980s, described by Watts as characterized by 'richness, diversity and comparative scope' making a 'profound impact' on the the debate, therefore, is the work of Byres, whose ideas about paths of transition informing discourse on the agrarian question are said to display 'a deep historical sensibility'.⁴⁰ In keeping with such encomia, Bernstein is credited by Akram-Lodhi, Dietz, Engels, and McKay with establishing 'more open and pluralist lines of enquiry', by van der Ploeg with an 'excellent' interpretation of the agrarian question, and by Pattenden for his 'politically significant' interpretation.⁴¹ Yeros is critical of

---

39   See Bernstein (2021b: 26–27). Elsewhere, and against evidence to the contrary, Bernstein (2018) has argued that there is no such thing as rich peasants, only middle ones.
40   Watts (2021: 59–60).
41   For these endorsements, see Akram-Lodhi, Dietz, Engels, and McKay (2021b: 4), van der Ploeg (2021: 109), and Pattenden (2021: 94). Misinterpreting yet at the same time privileging the views of Bernstein in this manner has precedents, on which see Brass (2007, 2013). In an earlier and similar tome – *The Elgar Companion to Marxist Economics* – complex and wide-ranging debate concerning the agrarian question was reduced to exchanges just between him and Byres, while in two other edited volumes – *Reclaiming the Land* and *A Radical History of Development Studies* – he was erroneously credited with having formulated virtually all of the most important theoretical insights in the field of agrarian studies.

Bernstein, scathingly describing his attempted reinterpretation of the agrarian question as one that 'almost negates the rich heritage of Marxist scholarship at one stroke', a welcome break with his own previous uncritical endorsement of Bernstein's views.[42] Nevertheless, in arguing incorrectly that Marxist theory is simply about smallholder 'liberation' and 'autonomy' suggests that Yeros – like van der Ploeg who uses the same concept – appears to see the agrarian question through populist lens.[43]

Reprivileging Agrarian Populism

Significantly, when reviewing another and similar handbook, Bernstein himself observes that '[t]he benefits of such pluralism, especially for teaching and learning, are that students can decide for themselves between alternative 'theorization(s) and normative positioning(s)' if they are presented with a substantial degree of coherence and lucidity, which this Handbook does not provide'.[44] With this assessment one can wholeheartedly agree. Ironically, and advocacy by Bernstein of pluralism notwithstanding, it is precisely this kind of 'alternative theorization(s) and normative positioning(s)', involving Marxist critiques both of his own concepts ('classes of labour') and of his agrarian populism, that are largely absent from the CAS *Handbook*. Most revealing, therefore, is the disparity that emerges simply as a result of the number of times an author is cited in the *Handbook*, as indicated by a quick count.

On the one hand, there are those (in the 'heroic' category) who feature prominently: Bernstein is cited 91 times, Akram-Lodhi 89, Borras 85, Edelman 66, Watts 40, James Scott 38, Byres 36, Scoones 29, and Desmarais 28. On the other are those (in the 'marginal' category) whose work merits scarcely a mention: Petras is cited only 10 times, Brass a mere 6, Curwen and Meera Nanda 2, and Raju Das not at all. This already large gap is compounded by the way ideas are themselves discussed. Hence those in the former category, cited anyway in large numbers, also have their views considered at length. By contrast, those in the marginal category, barely referenced in terms of citation, appear solely as names in lists featuring others, with little or no space given to the views held. The unfortunate impression conveyed by this differing emphasis is hard to avoid. One could be forgiven for thinking, therefore, that apart from the ideas and interpretations of those in the heroic category, nothing else of

---

42 Jha and Yeros (2021: 337).
43 Jha and Yeros (2021: 338), van der Ploeg (2021: 114).
44 See Bernstein (2022).

much relevance to the field of agrarian studies is worth reading, and certainly not anything produced by those in the marginal category.[45]

An immediate and obvious difference between these same categories is that those in the marginal group are largely associated with the pre-2009 JPS, while most of those in the heroic group are on the editorial board either of JOAC or of the post-2008 JPS. This divide is crucial, since it marks a political disjuncture, whereby the central role allocated hitherto to Marxist approaches was replaced by emphasis given to agrarian populist ones.[46] The same difference in emphasis is found in the *Handbook*, where positive references to agrarian populist interpretations by those of heroic status constitute endorsements, while the paucity or non-existent references to critiques by those deemed marginal suggest an absence of the necessity for endorsements.

Hence the differences which structure the heroic/marginal divide are ones that are politically significant. Whereas most of those in the heroic category (Bernstein, Borras, Edelman, Scott, Desmarais, Scoones) regard agrarian populism as empowering, progressive, and positive, those in the marginal category (Petras, Brass, Das, Nanda) are for the most part critics of agrarian populism, regarded by them as negative, conservative (or reactionary) and disempowering of class struggle. It comes as no surprise that the same gulf arises in the case of positive/negative interpretations about the viability of peasant economy, and its desirability. Unlike the approach of the marginal category, for whom smallholders are differentiated by class, the heroic category tends to see the peasantry as undifferentiated in terms of class.

An additional and relevant question concerns the extent to which the same imbalance between supporters and opponents of agrarian populism is reflected in the way the history of the two main journals associated with CAS is recounted, both in the *Handbook* itself and in the journals.[47] Hence the problematic narrative contained in the introductory chapter of the *Handbook*,

---

45 Not the least significant reason for regarding such an impression as misleading is the omission of any consideration given the important contribution to Marxist theory by Das (2014; 2017).

46 Although articles supportive of agrarian populism appeared in the pre-2009 JPS, throughout this period most contributions were informed by a Marxist framework. After 2009, however, this pattern has been reversed, noticeably so.

47 Broadly speaking, accounts of JPS history leave much to be desired. Thus, for example, the claim by Bernstein (2021a: 45) that 'the publisher Taylor and Francis, which had bought the publishing house [of Frank Cass] after its founder's death', is quite simply wrong. Frank Cass sold the JPS and his other journals to Taylor and Francis in 2003, four years *before* he died in 2007. It is not the case, therefore, that Taylor and Francis acquired the JPS only after his death.

whereby those editing the JPS moves seamlessly from Byres (1973–2000) and Bernstein (1985–2000) to Borras (2009–2023), omitting to mention Brass (1990–2008), the same partial version appearing elsewhere in the *Handbook*.[48] Missing from this history, therefore, is the presence of the person who – after Byres – happens to be the next longest serving editor of the JPS.[49] In part, such absences can be explained by events at the JPS in the period 1996–2008, involving changes to its editorship, to its editorial board, and to its political direction.[50]

The veracity of these criticisms, made initially during 2022, was borne out shortly after by the appearance of the JPS Flagship 50th anniversary issue (Vol. 50, No. 2, 2023), not the least significant aspect of which is the viewpoint it reflects, together with the identity of the contributors. All the processes and trends identified in that review article are not merely confirmed but writ large in the issue of the journal.[51] In a sense, therefore, it is fitting that Bernstein is the only one of the surviving editors from the 1973–2008 period contributing to this 50th anniversary issue of the JPS, since it underlines what has been known all along: that he is indeed an agrarian populist.[52]

Amidst a number of vainglorious utterances ('we as Editors of JPS are humbled and somewhat awed by the responsibility of holding this precious space and resource [and] look forward to robust and critical engagement across these platforms and spaces with the global JPS community'), a self-congratulatory but politics-lite and theory-lite narrative celebrates what is now an undisguised agrarian populist approach.[53] Rarely has the accuracy of a review article been

---

[48] For instances of this see, for example, Akram-Lodhi, Dietz, Engels, and McKay (2021b: 4), Friedmann (2021: 20), and Veltmeyer (2021: 596).

[49] It is difficult not to see this absence as symptomatic, given that references elsewhere to the history of the JPS also make the same error. Hence the statement (Bernstein, 2021a: 45) that '[t]ogether, Bernstein and Byres edited JPS until 2000', a depiction that avoids mentioning the fact that 'together' excludes the presence throughout a major portion of this period (1990–98) of a third editor, Brass, who then and afterwards (2000–2008) contributed substantially to the work of editing the journal. This is not the only such instance, as similar references to the JPS being edited only by two of its editors can also be found in Capps and Campling (2016), Bernstein, Friedmann, van der Ploeg, Shanin, and White (2018: 690, 699), Bernstein (2010: X; 2013: 162; 2021: 45), and Bernstein and Byres (2008: IX–XII).

[50] On these changes, see Brass (2005, 2015a).

[51] The review article in question appeared in *Critical Sociology*, volume 49, issue 3 (2023).

[52] See Bernstein (2023) which, since it contained a serious misrepresentation of JPS history, one paragraph of which was subsequently withdrawn by the publisher. It is this revised version that has now appeared, bearing on the first page a footnote announcing that 'This article has been corrected with minor changes'.

[53] Hall, Grajales, Jacobs, Sauer, Galvin, and Shattuck (2023: 447).

confirmed so soon after its publication. There it was argued that under the guise of belonging to an earlier peasant studies framework, Marxism had effectively been declared redundant by current JPS exponents of Critical Agrarian Studies (CAS), and earlier Marxist critiques of positions now championed by CAS were as a consequence either downplayed, misrepresented, or simply ignored. Populist approaches flourish, so much so that at times contributions to the post-2008 JPS read like the special pleading by a claque eulogising this kind of framework.

Equating the JPS with the Via Campesina, as does the 50th anniversary issue, underlines precisely the extent to which the journal has in the period from 2009 onwards become essentially a mouthpiece for agrarian populist ideas and views, to the virtual exclusion of its erstwhile Marxist approach.[54] This is evident, for instance, in the problematic manner in which one article frames what is termed the 'gilded age of agrarian political economy'.[55] Despite the fact that this period is said to cover only the years 1950–1985, it lasted well beyond that, at least up to and including 2008. In short, and where material published in the JPS is concerned, almost a quarter century after its purported demise. As problematic is that among the names appearing in this category 'gilded age of agrarian political economy' are those who have either never contributed articles to the JPS or have done so only much later, after 2008. By contrast, Marxists who have indeed contributed many articles to the journal over the 1973–2008 period – that is, to the 'gilded age of agrarian political economy' – are simply not mentioned. *Plus ça change.*

Conclusion

As one reads through the essays contained in the *Handbook*, amidst a jumble of problematic arguments and one-sided presentations, a pattern begins to emerge. Under the guise of belonging to an earlier peasant studies framework, Marxism is declared passé, and its critiques of positions now championed by CAS are either downplayed, misrepresented, or simply ignored. However, agrarian populism flourishes, so much so that at times contributions read like special pleading on its behalf. Consequently, it is hard to avoid seeing the result as a thinly-disguised – but unsuccessful – attempt to establish a two-fold hegemony: over what constitutes the field of study, and of who has contributed

---

54 See Hall, Grajales, Jacobs, Sauer, Galvin, and Shattuck (2023).
55 Borras (2023: 452).

to this endeavour. Regardless of whether or not this was the intention, and to use one of its favourite terms, CAS amounts in effect to a land-grab by exponents of populist interpretations.

Where development studies are concerned, the political significance of this epistemological shift is easily discerned. It surfaces in the erroneous claim made by CAS to have superseded previous analyses of the peasantry, a consequence it is said of having discovered a 'new' and more accurate development paradigm, replacing the agrarian question of Marxism with the agrarian myth of populism. A case, perhaps, of petty commodity producers being essentialized as 'natural', and thus remaining eternally as such, peasants can never become other than they are. Declared redundant by CAS is not just Marxist theory but also the non-Marxist modernisation framework, neither of which is addressed in any detail. Among the arguments misinterpreted as a result is the nature of class, the similarity between the views of Kautsky and Lenin on the agrarian question, and the difference between proletarianization and deproletarianisation, together with their respective effects.

Claims by CAS to the 'newness' of its approach notwithstanding, the agrarian populism informing this paradigm not only negates the very idea of development but also (and consequently) returns to the pre-1960s idealised perceptions of rural society. In effect, CAS reprivileges age-old images of traditional and custom-bound smallholding family farming that is neither differentiated nor subject to fundamental change, an interpretation that historically is central to peasant essentialism. This recuperation of agrarian populist discourse has itself been facilitated by a process of journal capture, supportive of academically fashionable postmodern/post-development theory. The latter was itself part of the wider post-1980s swing towards conservativism, a move unsurprisingly reflected in academic publishing, a transformation accounting in part for the problematic manner in which identity politics are perceived as 'natural' and empowering, and peasant economy together with the dynamic governing its reproduction are depicted by agrarian populism as viable. How all this plays out in terms of individual contributions to development theory is explored in the chapters which follow.

PART 2
Alternative Agendas

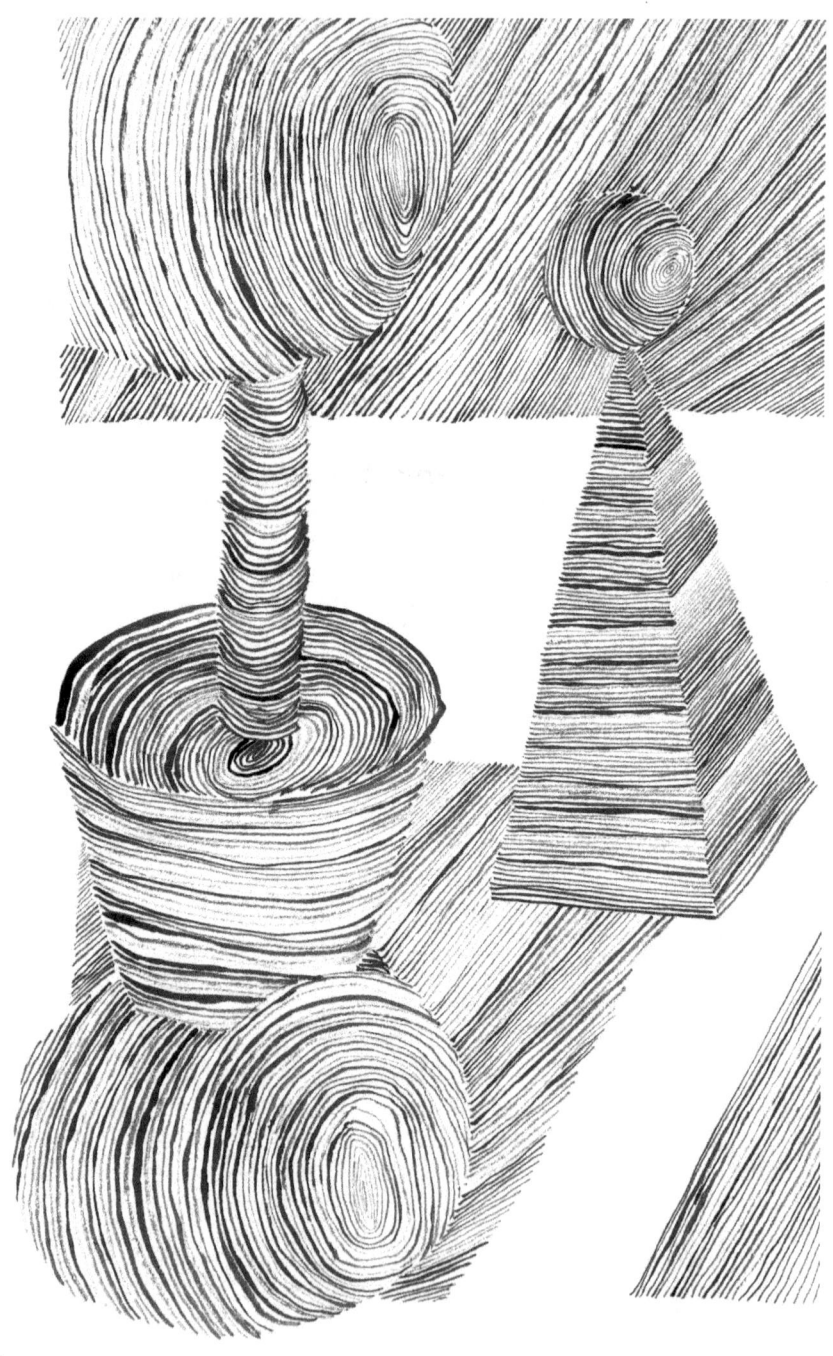

CHAPTER 5

Development: a Theory without a Past, Present, or Future?

> When I sit and warm my hands, as best I may, at the little heap of embers that is now Political Economy, I cannot but contrast its dying glow with the generous blaze of the vainglorious and triumphant science that it once was. Such is the distinctive character of modern learning, imprint with a resigned agnosticism, able to refute everything and to believe nothing, and leaving its once earnest devotees stranded upon the arid sands
>
> > An observation made over a century ago by STEPHEN LEACOCK, political scientist and humourist, in 'The Apology of a Professor: An Essay on Modern Learning', that still resonates.[1]

∴

Introduction: Paradigms/Concepts That Disappear/Reappear

Where systemic transcendence is concerned, one way or another the approaches to development-as-modernity considered in each part of this volume all sign up to more of the same, thereby depriving this discourse of its progressive aspect. The absence of the latter has serious consequences for any endeavour to address labour historiography, particularly since from the 1980s onwards the development model followed by many Third World countries has undergone a significant change. Instead of endogenous growth, the object has become to seek employment in metropolitan capitalist nations, where jobs already exist and are better paid. To use the old trope, development nowadays entails not fight but flight. Marxism, with its focus on the formation/consciousness/struggle of class, together with an emphasis on important development issues such as the industrial reserve and the desirability of

---

1 Leacock (1916: 26).

a transition to socialism, currently is the only theory which pursues an economic and political development that is designed to transcend capitalism.

Unable any longer adequately to address the interrelated pattern of economic growth, plus social and political change, therefore, those in departments of development studies located in western academic institutions have for some time now been engaged in a somewhat paradoxical enterprise. In effect, negating their *raison d'être*, by defining their subject (and ultimately themselves) out of existence. For those who study development, therefore, the choice is currently between a 'nicer' capitalism (to which society must return) and the denial that there is any longer such thing as capitalism (whatever exists cannot be transcended). In short, the object of development studies has changed: instead of moving beyond capitalism to a more progressive systemic mode, the only alternatives are either going back to a previous version of the same mode or succumbing to aporia, a form of intellectual stasis where any future model of change is to be addressed.

Accordingly, examined critically in this chapter are recent contributions made to this sort of discourse by two labour historians, van der Linden and Lucassen, each of whom subscribe to what they both claim is a new paradigm – global labour history (hereafter GLH) – formulated and applied by them to the study of development on a world scale. Challenging Marxist theory about twentieth century labour movements, van der Linden maintains that 'the continuous expansion of the working class [meant] that labour would become ever more important in future revolutions, which would result in the subversion of capitalist relations. This idea appears to have been refuted.'[2] Similarly, when he denies that capitalism developed first in England, van der Linden aligns himself with the currently fashionable revisionist view that capitalism as a system can be found everywhere at all times.

The corollary of this view is that, as a systemic form, capitalism is eternal, and thus cannot lead to anything that is different from itself. In other words, no transcendence, and therefore no systemic development leading to a different and more progressive mode of production, is possible. Of additional significance in this regard is that earlier, when charged with trying to jettison Marxist political economy, van der Linden strongly denied this.[3] Yet the Foreword to his latest book concedes this very point, observing that because he regards it

---

2 van der Linden (2023: 14).
3 For the charge, see Brass (2017a); for the denial, see van der Linden and Roth (2018).

as mistaken van der Linden in effect discards the development approach of Marx and Engels.[4]

What follows is divided into two sections, the first of which examines claims about revising our understanding of capitalism and its workforce, together with the implications of this for development studies. A critique of this approach, illustrated with reference to the lumpenproletariat category, is elaborated in the second section.

I

Call a Friend

Currently it seems almost *de rigueur* that any book on the subject of global development must either be preceded by praise lavished on contributors by an editor, or carry a foreword by a colleague/friend of the author making extravagant claims on the latter's behalf. The 2023 volume by van der Linden is no exception: in the foreword he is described effusively as not just 'the world's leading labour historian' because he 'rethinks the global working class from the ground up', but is also responsible for asking 'the field's biggest questions such as the definition of labour under capitalism' and how to think about 'different kinds of coerced labour'.[5] In much the same adulatory vein, we are informed that the essays by van der Linden constitute 'an entirely new logic ... needed to rejuvenate the academic field of labour history and reorient our thinking about global capitalism', and further that the purpose of the essays is to 'emphasize the need to define causal mechanisms instead of building increasingly general theories', thereby 'challenging some of the holy cows of the social sciences'. Not a single one of these claims, forming as they do a litany of exaggeration, is correct.

That van der Linden is 'the world's leading labour historian' is contentious, and no more than hyperbole, particularly when such a claim on his behalf is considered in relation to the list of other candidates for that title.[6]

---

4   In the foreword the relevance of Marx and Engels to the approach of van der Linden (2023: XIX) is dismissed peremptorily: 'Their view of working-class formation: wrong. Their view of the source of value under capitalism: wrong. Their view of the non-European world: wrong. Of revolution: wrong'.

5   The foreword is by Sven Beckert, who is also mentioned favourably elsewhere in the text (van der Linden, 2023: XVII–XX, 8, 15, 16, 62, 68).

6   A list of those whose claim to pre-eminence as 'the world's leading labour historian' surpasses that of van der Linden is very long, and includes among many others important scholars like Eric Wolf, E.P. Thompson, Dorothy Thompson, Philip Foner, David Montgomery, Herbert Gutman, Peter Linebaugh, David Roediger, Dick Geary, Barbara

As problematic is the claim that he 'rethinks the global working class from the ground up', since unlike others writing about labour history across time and space, van der Linden has never undertaken fieldwork in Third world countries, and therefore lacks the first-hand knowledge of conditions and relational forms at the rural grassroots in contexts on which comparative analysis – the global dimension of development studies – depends.[7] In keeping with this are questionable claims made by van der Linden himself about the novelty of GLH, together with the pathbreaking insights he thinks it can deliver.[8]

Hence the emergence of GLH in around 2000 was preceded by studies of the new international division of labour, an approach the focus of which was on the way capital restructured the labour process by outsourcing manufacturing and employment to locations in the developing countries where labour-power was less costly and available, a process licensed by a recent expansion in the world-wide reserve army of labour.[9] Long before that, even, Marx, Lenin, Trotsky, and Rosa Luxemburg all made much the same point, as did subsequently C.L.R. James, Eric Williams, and others. Perhaps because of this necessity to formulate a totalizing approach, one that is applicable everywhere, GLH is defined by van der Linden in vague, catch-all terms. In effect it is a concept so general and broad in coverage that nothing appears to exist outside it – a history of everything, everywhere, at all times.[10]

Solow, Robert Fogel and Stanley Engerman. To this same list might be added other scholars like Immanuel Wallerstein, Fernand Braudel, Peter Worsley, and James Petras. Whatever its final composition, it is hard to see how the list might be said to include van der Linden.

7   Given their methodological approach, anthropologists have an evident advantage in this regard, since they are able to study labour relations directly at village, household, or kin group level. The kind of misinterpretation and errors that arise in the case of historians writing about the rural Third World without having conducted similarly extensive participant/observation there are evident in the work of Hobsbawm about Latin America (on which see Brass, 2018a: Chapter 9).

8   'I consider global labour history to be a distinctive field of research', observes van der Linden (2023: 5), 'just like art history or linguistics'. Just how such an approach can be said to be different from the wider analysis that is development studies – how, in short, is it possible to consider labour abstracted from all the other issues constitutive of political economy (industrial growth, agrarian transformation, state formation and reproduction) – is a question that is not raised.

9   See in particular the seminal text by Fröbel, Heinrichs and Kreye (1980).

10 GLH is defined (van der Linden, 2023: 6) as 'primarily concerned with describing and explaining the intensifying (or weakening) connections (interactions, influences, transfers) between different world regions, as well as the economic, political, social, and cultural networks, institutions and media have played a role'. Despite his insistence that GLH 'is not a "history of everything"', it is clear from the way he defines it that this is precisely what it claims to be.

Not the least problematic aspect of this wide-ranging, amorphous sort of approach is its methodology. Although he accepts that comparative studies 'exploring the causes and consequences' of development 'are an integral part of global labour history', not addressed is – in the absence of fieldwork – how epistemologically to reconcile (let alone synthesize) different studies informed by dissimilar theoretical approaches, methodologies, and research agendas.[11] Of similar concern is the significance attached by van der Linden to compiling data sets, replicating thereby the same claim to methodological novelty on which cliometric historiography of the mid-1970s based its reinterpretation of plantation slavery in the antebellum southern states, a methodology subsequently revealed to be flawed.[12]

Nor is it the case, as the foreword asserts, that since it includes slaves, indentured labour and women, the extended concept of working class used by van der Linden is new, and a break with past theory. Marxism has always included these same categories in its analyses, but *not* as belonging to the proletariat. Hence the inference that slaves, indentured labour and women are (and always have been) consciously excluded by Marxist historians and social scientists or are merely and unaccountably absent from these analyses is self-evidently wrong. Equally contentious is the view that van der Linden has formulated how we think about coerced labour: not only is he a latecomer to the long-standing discussion about this issue, but – as will be seen below – he fails to understand what characterizes labour-power as unfree. This in turn can be attributed to the privileging by him of empiricism over theory, an approach that has been described as a form of butterfly collecting.[13] In the end, book

---

11   See van der Linden (2023: 6).
12   The attempt by cliometric historiography to revise the meaning of slavery is associated with the work of Fogel and Engerman (1974); the powerful critique of the methodology informing this approach was made by Sutch (1975).
13   For this description, see Brass (2018a: 149ff.). That the approach of van der Linden is empiricist is conceded in the foreword (van der Linden, 2023: XVI). There is an unmissable irony here. Privileging empiricism over theory, elsewhere van der Linden (2016) asserted that it was necessary to look at all the empirical evidence – 'all forms of coerced labour' – in minute detail (for a modified version of the same argument, see van der Linden, 2023: 84ff.). However, in the end he was compelled to concede that attempting to identify multiple variants always returned to a basic dichotomy – the free/unfree distinction – long known about and underlying the many forms composing his taxonomies. In other words, the empiricist categorization of production relations ends up conceding what it starts out by denying: namely, that what structures both the definition and effects of all the variants identified is the simple polarity entry-into/exit-from the work arrangements concerned, a dichotomy which – as Marxist theory about deproletarianization has always argued – in turn informs the difference between labour-power which is free and that which is unfree.

prefaces/forewords carrying endorsements by friends/colleagues of the author possess the same amount of credibility as a television commercial for a supermarket product or a broadcast on behalf of a political party, and should be treated as such.

Concepts, Origins

Hyperbole provided by someone else is frequently accompanied by another sort of hyperbole, the product of the self. Hence the external hyperbole encountered in the foreword of his 2023 book can be linked in turn to (and is possibly an effect of) its internal variant, in the shape of problematic claims made by van der Linden himself concerning the originality of concepts deployed or the case made. Like a number of other social scientists, therefore, he tends to present as a 'new' discovery – and then attribute it to the GLH approach – issues to do with the nature of the labour regime that have not just been known about for a very long time, but also much discussed. Discoveries perceived as 'new' are said to be 'revealed' by what follows in the text, or signalled even more directly under the rubric of a 'new concept' of his own plus 'first research results' incorrectly attributed to GLH.[14]

Observing that forms of unfree labour-power (convicts, debt bondage) 'are now also of considerable interest', the accompanying endnote lists mostly GLH exponents, thereby conveying the misleading impression that such interest is recent and the result only or largely of GLH research.[15] As problematic is the

---

14   For these see van der Linden (2023: 6, 79). Part of the difficulty facing him is that by attributing the case about unfree labour to many different authors, most of whom are GLH exponents, the impression conveyed is that he is merely synthesizing elements from different sources on which to base his case, this rather than citing a single non-GLH source where the same kind of arguments were originally made. Boosting GLH in this manner, although understandable when seen from his point of view, cannot but generate a false sense of the intellectual genealogy, resulting in a one-sided account of debate about the issue in question.

15   See van der Linden (2023: 6–7, 15). In keeping with the privileging of GLH, elsewhere in the book (van der Linden, 2023: 62 note 22, 68) the argument that 'commodification of human labour ... can also be based on physical coercion, as is the case with indentured labour or chattel slavery', is sourced to one of his own earlier publications. Noting that 'research in recent years has revealed that many so-called free workers were really *bonded* labourers', van der Linden (2023: 135, original emphasis) also omits to mention that this very point has been made for many decades now in relation to Latin America, India (contra Breman and others), and ancient society (contra Banaji), by those unconnected with GLH.

claim to having established 'a new typology' where the characteristics of unfree labour are concerned.[16] The importance of the combination entry-into + exit-from for the definition of unfree labour – central to that 'new typology' – was actually invoked elsewhere back in 2003, against Banaji and neoclassical economic historians, all of whom were using only the entry-into criterion as defining the unfree nature of labour-power.[17] Equally, 'the implicit assumption of the present-day notion of labour markets is that the workers have to offer their labour-power *themselves*' is not new either, being an issue raised in a debate – again – with a neoclassical economic historian some three decades ago.[18]

By announcing that the meaning of 'working class' is disputed, van der Linden signals an intention to extend the boundaries not just of its historical presence but also of its components.[19] That wage-labour has been 'performed … for thousands of years', plus the fact that the working class has 'its subjective side, manifested by its culture, mentality', leads inexorably to support for the view that '[i]n recent decades, more and more voices argue that the … interpretation of "working class" is too restrictive'. From this it is but a logical next step to subsequent claims: first, that 'there are all kinds of forms of "hidden" wage labour, such as sharecropping [and] self-employed workers'; and second, that 'this relativization of the boundaries of the working class has recently motivated historians to redefine it, such that slaves and other unfree workers can also be included.'

Such relativization/redefinition depends in turn on working class membership being conceptualized in terms not of production relations – as it is for Marxism – but rather and simply as labour which is commodified. Observing that '[c]ommodified labour-power does not solely comprise "free" wage labourers', van der Linden then argues that 'commodification of labour-power has many manifestations, including chattel slavery, sharecropping, or debt peonage'.[20] For him, therefore, '"free" wage labourers are no different from

---

16   According to van der Linden (2023: 12), therefore, 'I identify three essential aspects (entry into the labour relation; extraction of labour during the labour relation; and exit from the labour relation) and, in the process, endeavour to establish the foundation for a new typology'.
17   The necessity to unfree production relations of the entry-into + exit-from combination was an important feature of the critique by Brass (2003: 117) of Banaji (2003) and Steinfeld (2001).
18   See the exchange between Brass and Schlomowitz during 1990–91 in the journal *Slavery and Abolition* (11/1, 12/3). The same issue was also referenced by Brass in a contribution to a volume on unfree labour edited together with van der Linden (Brass and van der Linden, 1997).
19   For this and what follows, see van der Linden (2023: 70ff.).
20   van der Linden (2023: 12).

other labourers in being unfree from coercion.' As will be seen below, this all-encompassing concept of belonging to his notion of an 'extended working class' is true also of the way the lumpenproletariat is perceived.[21]

Capitalism Everywhere, Capitalism Nowhere

Given the extended boundaries of what van der Linden understands as the working class, it is perhaps unsurprising that capitalism, too, is reconceptualized by him as possessing a similarly enlarged presence in terms of time and space.[22] As to the meaning of capitalism, what it is, where and when it exists, it is clear from his comments that van der Linden belongs to the category of those who maintain that capitalism cannot be transcended. Asking the question 'will capitalism also reach an endpoint?', he answers: 'Marxists ... have often claimed that capitalism is doomed in the long run [but others have] rightly pointed out that "capitalism has never really been threatened by collapse ... in the last 150 years."'[23] For those like van der Linden, who see capitalism in all periods of history and in all global spaces, therefore, their syllogistic reasoning appears to operate in the following manner. Capitalism can and does use unfree production relations, and as these forms of labour are found in many contexts before industrial capitalism emerged from Europe, therefore this economic system itself not only pre-dated the latter development but is also found earlier in non-European (Asian, Latin American) places.

This revision mimics that pursued during the 1970s by cliometric historians like Fogel and Engerman.[24] In each case, capitalism as a systemic form is depicted as an historically eternal mode, ever-present and incapable of carrying within itself the seeds of its own demise: consequently, development not

---

21 The claim that no difference exists between the lumpenproletariat and the working class is not new either, having been made in the recent past by postmodernists such as Foucault (1996) and Žižek (2014).
22 That van der Linden uses an extended notion of capitalism, as regards both space and time, is evident from many of his comments. Hence the observation (van der Linden, 2023: 11, original emphasis) that '[i]n my view, we will also need to reinterpret the history of the last few centuries – possibly revealing ... that England was *not* the cradle of industrial capitalism'. Further on, he (van der Linden, 2023: 64, 66) accepts that a 'broad definition of capitalism will inevitably lead to a different periodization than a narrow one', adding: 'A drawback of many typologies and periodizations of capitalism is that they are based on the histories of the old core regions in the world-system: Western Europe, North America, and Japan'.
23 See van der Linden (2023: 67, original emphasis).
24 See Fogel and Engerman (1974).

merely leading to but necessitating a systemic transformation vanishes from the agenda. In order to arrive at the same conclusion – that the economic process concerned was a capitalist one – it was necessary for them to redefine its unfree production relation (chattel slavery) as a form of freedom. This in turn enabled the antebellum southern plantation itself to be presented as capitalist, on the grounds that its workforce consisted of labour that was not unfree.

Invoking another cliometrician – this time Steinfeld – as a supporter of the view that because free workers are bonded by debt, consequently no difference exists between free and unfree labour-power, van der Linden is seemingly unaware of the difficulties with this argument, and what a neoclassical economist concludes from it.[25] For Steinfeld, therefore, the sameness derives from the fact (for him) that both free and unfree 'possess the same ability to choose to work'.[26] Not only does this difference-dissolving claim overlook abundant evidence to the contrary – slaves are unable to choose whether or not to work – but to regard both free and unfree labour as 'choice-making' is consistent with the idea of 'subjective preferences', a concept at the heart of neoclassical economic theory. In short, for the latter approach there is no such thing as unfree labour, since all workers are by definition free to choose their employment, a view that is very different from the interpretation held by van der Linden himself, which is that the similarity derives from their being both subject to coercion: that is, each is unfree.

Where the history and presence of capitalism itself is concerned, this sort of epistemological realignment has generated claims that can only be described as odd. Thus, for example, some now argue that as capitalism has in effect never existed, there is no such thing as capitalism in the sense Marx and Marxism understood. Having indicated that he is not 'against Marx', therefore, Lucassen announces that he intends to compose a history of work stripped of concepts not just like class and class struggle but also like capitalism and modernity. This strange methodology is itself compounded by his intention, nevertheless, to retain terms such as market, labour relations, social inequality, collective action and exploitation.[27] Quite how the latter set of concepts are

---

25   What follows was outlined initially in Brass (2011: 113–14).
26   Steinfeld (2001: 318).
27   Because 'the central concepts of capitalism and modernity are now in flux and thus have lost their precision and original analytical function', argues Lucassen (2022: XV–XVI), it 'poses a problem for writing a long-term history like this'. He continues: 'For this reason, I have refrained from giving the terms *capitalism* (and the associated *class* and *class struggle*) and *modern* (versus *traditional*) a central place in this book [since] I believe that these terms have become so contaminated in the discussions of the last [100–150 years] that they have lost their analytical power in global labour history (original emphasis)'.

to be defined and applied once they are decoupled epistemologically from the former group, is never revealed. Unsurprisingly, this approach generates yet more confusion and contradiction.

Initially, therefore, Lucassen opposed categorizing unfree labour as a residue of a pre-capitalist social order, declaring unambiguously that understanding the accumulation/unfreedom link depended in turn on how one defined capitalism itself.[28] A decade later, however, his position has shifted to one in which he denied that unfree production relations continued or increased following the abolition of slavery ('I resist the impression which is sometimes created that unfree labour since the abolition of slavery continued, or even increased, until the present').[29] Yet another decade passes, and Lucassen is now compelled to acknowledge that unfree labour does indeed persist into the present, a puzzlement expressed by the constant reference to its 'tenacious' reproductive capacity.[30]

The Marxist view based conceptually on deproletarianization is different from these revisionist approaches, not least in its focus on the decommodification of labour-power that is the sole property of the subject as distinct from its commodification personally by its owner, the worker.[31] Unlike other interpretations considered here, therefore, it historicizes unfreedom, tracing

---

Instead, 'I do, however, wield [sic] the terms that are behind them, in particular *market ... labour relations, social inequality, collective action* and ... *exploitation*'. This is rather like announcing an intention to swim the channel and then declaring it unnecessary first to know how to swim.

28   Critical of the view that unfree production relations 'were fundamentally noncapitalist', Lucassen (1993: 17–18) argued strongly that the contrary was the case: 'All depends on the definition of capitalism of course. As we may be sure by now there is a strong relation between the occurrence of international economic development and unfree labour ... and that consequently unfree labour cannot be seen as a phenomenon, restricted to some primitive or backward situations'.
29   Lucassen (2013: 14 n47).
30   Lucassen (2022: 322, 323, 360).
31   See Brass (1999, 2011) for deproletarianisation, a concept based on Marxist theory about class struggle. It refers to the process whereby agribusiness enterprises, commercial farmers and rich peasants reproduce, introduce or reintroduce unfree production relations. Such workforce restructuring involves replacing free labour with unfree equivalents, a procedure frequently resorted to by employers. The object of this labour process decomposition/recomposition is to discipline and cheapen labour-power, an undeniable economic advantage in a global context where capitalist producers had to become increasingly cost-conscious in order to remain competitive. Following on from the adoption of the Green Revolution in the Third World, coupled with labour market deregulation in metropolitan capitalist nations, such workforce restructuring has benefitted from the capacity of producers to draw on a burgeoning industrial reserve army.

the systemic dynamic whereby initially free labour can – and does – become an unfree equivalent. Arguing that once established in metropolitan contexts, competition between capitalists resulted in increasingly acute forms of class struggle waged 'from above', leading to the resort by producers to labour-power that, because it was unfree, was cheaper and easier to control: in short, depriving the worker who owns this commodity of the ability personally to sell it to another purchaser. It is the loss of this capacity that constitutes deproletarianization.

Furthermore, when found in Third World plantations, such unfree production relations are not evidence for a peripheral or local capitalism, as inferred by the capitalism-everywhere approach, but rather capitalism developed within and operated from (and in the interests of) the systemic core: that is, metropolitan capitalism. Those who maintain such instances constitute evidence for the emergence of a non-European capitalism, an autonomous systemic dynamic that not only originated but was also reproduced solely in the non-metropolitan context involved, are mistaken. There was trade, certainly, but as all Marxists (and most non-Marxists) know, this does not of itself equate with the development of capitalism as an economic system.

II

It is unfortunately still the case that some leftists appear to have difficulty in recalling that for Marxist theory the element of class is not simply an add-on, to be considered once victimhood has been conferred on all other identities. Class is, and has always been, for Marxism the master (or 'foundational') identity, one which drives the process of historical change and its specific patterns of systemic transition. Hence abandonment of class as the master narrative, and its replacement by non-class identities, leads in turn to a variety of politically symptomatic transformations. What is deemed to be desirable as 'from below' agency and its objective, therefore, is recast: from revolution to resistance; from state capture to state avoidance; and from socialism to a 'nicer'/'kinder' form of capitalism. Needless to say, this kind of shift cannot but have profound implications for the study of development, its purpose and objective, together with its desirability and beneficiaries.

Development Theory?

Where Lucassen is concerned, therefore, the problems a recognition that unfree production relations as 'tenacious' poses for his current book about the history of work are all too clear.[32] Hence the narrative about unfree labour, central to his account of work throughout history, is faced with the difficulty informing the attempt to connect – or disconnect – the presence of unfree labour to a systemic form – capitalism – which Lucassen has discarded. His earlier argument basing any discussion regarding the presence/absence of unfree labour on a prior understanding of capitalism as a system has been replaced: first, by aporia concerning when/where such a mode of production emerged; and latterly by abandoning the term altogether.[33] It is no longer possible, therefore, to pose the question of whether capitalist development entails an increase or decrease in such production relations. Having jettisoned the concepts enabling him to do this, and unable as a result to decide either what characteristics define a production relation as unfree or whether the latter is increasing, he cannot avoid contradiction.

Its central role in his story about work notwithstanding, what unfree labour is, and thus whether it is increasing, are issues which remain unclear. In the case of debt bondage, uncertainty as to its relational status is evident from the description by Lucassen as a 'credit-securing strategy', a labour relation which because it is entered into 'voluntarily' is, it is inferred, free.[34] This is no different from the way neoclassical economic historians misinterpret the relation, a similarity borne out when free wage labour is defined by him as the capacity on the part of an individual 'to conclude work contracts': that is, to enter a production relation voluntarily (= freely).[35] This is the same error as that made by cliometric historians and Banaji, who – as indicated above – similarly overlook the fact that the freedom of a work contract lies in the ability not just freely to enter but also freely to exit from the relation (entry-into + *exit-from*). In the same text, however, indentured labour working to pay off debt – an unfree production relation – is nevertheless downgraded by Lucassen to 'shades of unfreedom' (= *perhaps* an unfree relation), which hints at the equally problematic

---

32 See Lucassen, '"Modern Slavery": Why is Unfree Labour so Persistent?', International Institute of Social History blog, 16th November 2022.

33 For this conceptual aporia/abandonment where capitalism and modernity are concerned, see Lucassen (2022: XV–XVI).

34 See Lucassen (2022: 277), who underlines his view that debt bondage does not correspond to unfree labour by putting inverted commas round the accompanying word 'enslaved' (= not really unfree). In keeping with this, subsequent references (Lucassen, 2022: 116, 196–97, 449 n57) suggest either that only chattel slaves are unfree, or that neither sharecropping and debt bondage constitute unfree labour.

35 Lucassen (2022: 196–97).

argument that unfreedom is merely situated on a relational 'continuum', one of many locations thereon.[36]

On the question of whether there is an increase or decrease in the incidence of unfree labour, because he rejects concepts like class struggle, which links the way production relations change according to the flow and ebb of conflict, Lucassen views the shift from unfree to free labour as a unilinear transformation, not unlike whig historians.[37] This is compounded by other erroneous assessments. First, that an effect of colonial expansion was that a free workforce ('Western labour relations') was reserved for their own nations by Europeans, who nevertheless imposed unfreedom on colonized subjects.[38] Reproducing the world systems theory of the 1980s (free labour at the core, unfree labour at the periphery), Lucassen overlooks the widespread use of unfree production relations throughout metropolitan capitalist nations.[39] And second, the mistaken and linked assumption that unfree labour declines as the market expands, which not only reproduces the semi-feudal thesis argument that accumulation and unfree labour are incompatible, but overlooks much evidence to the contrary.[40] Unfree labour increases as the market spreads, because employers are forced to cut costs so as to remain competitive and survive as capitalists, an objective realized by restructuring the labour process where possible, so as to replace free workers with unfree equivalents.

Turning to the case made by van der Linden. the attempt by him conceptually to extend working class membership by means of relativization/ redefinition is faced with analogous methodological and theoretical difficulties. Methodological problems stem from the fact that van der Linden builds his whole case for GLH on the claim that hitherto analyses of economic and social development have lacked direct knowledge ('from the ground up') of

---

36 Lucassen (2022: 309).
37 As depicted by Lucassen (2022: 304, 362), the road of travel seems to be all in one direction, embodied in comments such as '[t]here has been an undeniable waning of unfree labour over the last two centuries', and '[d]ue to the global shifts in labour relations from unfree to free labour'.
38 Lucassen (2022: 194).
39 Also overlooked is that the presence in what became colonies of unfree labour long pre-dated European expansion into such contexts.
40 References by Lucassen (2022: 12, 191, 293) to the decline of unfree labour as the market expands include the following: 'Slowly but surely, the importance for the market of domestic, independent and [,] somewhat faster, unfree labour disappeared'; 'Deep monetization of the already existing market economies has proved crucial for the development of free labour … all working for the market'; and 'In the last two centuries, labour relations – now primarily market oriented shifted radically. The share of unfree labour fell sharply'.

rural contexts outside metropolitan capitalism, and have consequently missed crucial similarities and differences. Acknowledging the usefulness of anthropological participant/observation methodology, van der Linden nevertheless omits to consider the implications, both for his own research and for the wider GLH project, of the absence of such an approach.[41] When he observes that 'historical scholarship comprises both meaningful silences and misleading conceptualizations', van der Linden is unaware perhaps as to the full extent of its applicability, including to himself.[42]

Clearly, it is not absolutely necessary to possess fieldwork experience of a particular context in order to understand what occurs there in terms of labour regime, its causes and outcomes. The problem faced by van der Linden is that since GLH privileges direct knowledge of non-metropolitan contexts, in particular rural areas of Third World nations, he in effect makes an epistemological virtue of an understanding premised on the sort of direct knowledge of the grassroots which he himself does not have.[43] In cases where social scientists and historians have to rely mainly on secondary sources, moreover, it is crucial not just to know what kind of concepts and theory structure the information provided by these sources, but also to interrogate the latter. This, too, he fails to undertake.

Symptomatic of these difficulties is the attempt to justify the presence of the many errors and contradictions about unfree labour found in the work

---

41 See van der Linden (2023: 11). It is possible to consider the impact on the analysis by Deas (1977) of a coffee-growing estate in Colombia over the late nineteenth and early twentieth century in order to understand the kinds of methodological difficulty faced by historians. Based on correspondence between landowner and administrator held in the estate archives, the narrative reproduces glimpses of landlord/tenant conflict as seen 'from above'. How tenants and labourers perceived the situation, and why, is not depicted in the record. Hence the 'voice-from-below', and its account of these same events and struggles as presented not by the landlord and administrator but rather by peasants and workers themselves, can be said to be lacking.

42 van der Linden (2023: 10).

43 The perils of methodologically privileging direct grassroots knowledge of a place without actually having this are highlighted in the 1940s critique by Graham Greene (1970: 252–54) of a book by J.B. Trend, *Mexico: A New Spain with Old Friends*, the first paragraph of the review merits quoting at length: 'This is an account by a Cambridge professor of two trips to the tourist resorts of Mexico, but it pretends to be rather more. The professor is a Spanish scholar – and that should have been an advantage; but he was handicapped by the unenterprising nature of his journey …, by his friendships with Spanish Republicans as strange to the country as himself, by his ignorance of and antipathy to the religion of the country, and by a whimsical prose style less successful in conveying the atmosphere of Mexico than that of Cambridge in-jokes on the Trumpington Road, charades with undergraduates in red-brick villas, bicycles in the hall'.

of Jan Breman. On the latter's approach, therefore, van der Linden adopts an unsustainably benign interpretation of what are clearly mistakes or misunderstandings on Breman's part, classifying them instead in positive terms as 'a layered deployment of concepts'.[44] The result borders on the ludicrous: how, for instance, is it possible to regard as nothing more than 'a layered deployment of concepts' the difference between an initial argument by Breman (that unfree labour-power is incompatible with capitalism, and the spread of the latter necessarily entails the displacement of bonded labourers by free workers) with the subsequent and opposite claim made by him (that not only is unfree labour-power compatible with capitalism but in many instances is its preferred relational form)?[45]

Furthermore, as the reference to 'hidden' forms of wage-labour attests, van der Linden also falls into what might be called 'the Banaji trap', whereby the privileging of labour-power commodification leads to the mistaken conclusion that, since all its components are paid, they must in relational terms all be alike (= 'disguised wage-labour'): akin to saying that, because we all breathe air, we are all the same.[46] A difference exists, however, in that the interpretation of van der Linden is a reverse variant of the Banaji version: whereas in the case of the latter all who undertake paid work are *ipso facto* wage-labour that is free, for van der Linden by contrast all those in the same category are unfree. When van der Linden denies that capitalism developed first in England, he aligns himself with the currently fashionable revisionist argument that – together with unfree labour – capitalism as a system can be found everywhere at all times. The corollary of this view is that, as a systemic form, capitalism – like unfreedom – is eternal, and thus cannot lead to anything that is different from itself. In other words, no transcendence, and therefore no systemic development, is possible.

---

44   van der Linden (2023: 45).

45   For evidence of this *volte face* in the way Breman interprets the capitalism/unfreedom link, see Brass (2017b: Chapter 16) and Chapter 8 in this volume. Avoiding this change of mind, let alone the reason for it, van der Linden (2023: 44) seeks instead to justify the dissonance between earlier and later versions by arguing implausibly that Breman 'has been able to interpret the same phenomenon (such as bonded labour …) in different ways over the years without finding this problematic'.

46   Additional evidence suggests that van der Linden does indeed follow in the footsteps of Banaji. Not only does van der Linden reproduce the same argument as Banaji about 'hidden'/'disguised' wage labour, but the former has also provided an enthusiastic if inaccurate endorsement of the latter (Banaji, 2010: XI–XV), and more recently privileged the same commercial capitalism argument (van der Linden and Breman, 2020) as him.

Nevertheless, having labelled the nineteenth century Marxist concept of proletariat 'narrow', van der Linden then uses it as a springboard for his argument that 'we have to rethink the traditional notion of the working class', since in his opinion 'the process of class formation [in the Global South] was never completed'.[47] Not only is this quite simply wrong, but it is a claim that could only be made by someone who has never conducted fieldwork research at the rural grassroots. In the case of Peru, for example, research confirms that the presence and role of bonded labour in 1960s lowland Amazonia was itself part of the class struggle waged 'from above' by capitalist rich peasants, erstwhile estate tenants, better to control, cheapen, and retain the labour-power of landless migrants and/or poor peasant ex-sub-tenants. Research conducted in Northern India during the 1980s indicates that much the same kind of process occurred there as well.[48] That in these contexts non-class identities are reproduced 'from above', as a form of false consciousness, is misunderstood by van der Linden, because such phenomena can only be seen (literally) as a result of direct fieldwork experience in rural Third World countries.

So broad a definition of working class membership – based simply on commodification rather than production relations – also risks drawing into its net two different social categories, both of which are antagonistic to working class political interests. First, those from above: the higher echelons of the corporate world, who are not merely not working class but whose role in connection with the latter amounts to that of exploiters and oppressors.[49] And second, those below, in the shape of the lumpenproletariat. Although the earlier championing of the latter category by van der Linden, indicated by the positive references to the progressive and radical character of the lumpenproletariat as part not just of the working class but also its struggle, has diminished somewhat, he still adheres to the view that such elements should be included within the ranks of what he now terms the 'extended working class'. This despite – or

---

47   van der Linden (2023: 73).
48   These research findings from Peru and India, by no means the only ones coming to the same conclusion, are presented elsewhere (Brass, 1999: Chapters 2, 3, and 4).
49   The dangers of this spurious kind of equivalence can be illustrated easily, since this was precisely the claim made by the revisionist Arnold Bauer (1979: 41) in order to contest the coerciveness of debt peonage on the Chilean estate system. His argument was that no difference existed between on the one hand wealthy executives 'tied' to a corporation by bonus payments and pension schemes, and on the other impoverished agricultural labourers compelled to work for long hours in bad conditions in order to repay a loan taken from a landowner or labour contractor.

perhaps because – such a view is not consistent with Marxism.[50] The political difficulties with his benign view about the lumpenproletariat are hard to avoid.

The Sharpest Weapon

Sharing with the industrial reserve both location and many socio-economic characteristics, the lumpenproletariat does not – and cannot – fit the radical political image conveyed by van der Linden. Outside the formal economy, each of these two categories are drawn upon by owners of the means of production not just to replace existing workers who organize in pursuit of higher wages and better conditions, but also as strike breakers. The bargaining power of the lumpenproletariat is itself negligible, not least because lacking a common set of economic interests, it is composed of different elements which in class terms are unable to agree on a political programme, let alone unite around such an agenda. In part this is due to the rural origin of its components, since many of those who migrate to and settle in urban contexts, adding to the ranks of lumpenproletariat and becoming thereby components of the industrial reserve, are peasants displaced by capitalism.

Initial views depicted the lumpenproletariat in a negative light. Not just Marx and Engels, therefore, but also Lenin regarded this category as unambiguously reactionary, a counter-revolutionary force historically that in situations of political crisis sided not with a mobilized working class but rather against the latter and with the bourgeoisie.[51] From the development decade onwards, by contrast, a more benign interpretation emerged, as the lumpenproletariat was perceived in more positive terms. Now it was presented as composed of a deracinated peasantry that, having settled in the urban Third World slum or

---

50   Criticizing the anarchist Max Stirner for similarly conflating lumpenproletariat and working class, Marx (1976: 202, original emphasis) pointed out how for Stirner 'the entire proletariat consists of ruined bourgeois and ruined proletarians ... who have existed in every epoch', adding that consequently Stirner 'has exactly the same notion of the proletariat as the "good comfortable burghers"'. What vexed Marx was the mistake of 'identifying the proletariat with pauperism, whereas pauperism is the position only of the ruined proletariat, the lowest level to which the proletarian sinks who has become incapable of resisting the pressure of the bourgeoisie'. For Stirner, therefore, 'the lumpenproletariat becomes transformed into "workers"', whereas the latter for Marx 'in certain circumstances count as paupers but never as proletarians'.

51   Negative views about the lumpenproletariat, specifically its reactionary politics and strikebreaking role, are scattered throughout the writings of Lenin: see, for example, his *Collected Works* volume 5 (p.157), volume 15 (pp. 384–385), volume 18 (p. 161), and volume 26 (pp. 460, 468).

shanty-town, could and should be seen as an actually/potentially revolutionary force.⁵²

This revision stemmed from two interconnected ideas: that the metropolitan capitalist working class had been co-opted politically and economically, and was thus no longer revolutionary; and that in developing nations peasants forced off the land who migrated to urban slums were instead the new revolutionary subject, as such in the vanguard of the anti-imperialist/anti-capitalist struggle. In the absence of a revolutionary subject to be found in western capitalist nations, therefore, eyes turned towards the Third World in search of an alternative, and found it in the form of rural migrants newly arrived in urban areas. The latter, it was argued, composed a lumpenproletariat which, given its socio-economic characteristics (under-/unemployment, ethnic identity), would form the spearhead of future radical movements aimed at the overthrow of imperialism and capitalism, in the course of pursuing national liberation and socialism.

In his magisterial account of the mid-nineteenth century English working class, Engels outlined in the clearest possible terms the connection between on the one hand the presence of the lumpenproletariat, the role of the industrial reserve army, and their effect on class struggle, and on the other how capital used this particular combination to enhance its own profitability by means of increased labour market competition.⁵³ That the latter process has a deleterious impact on the wages and conditions of those in work is made equally clear by Engels, who charts the presence of a twofold competition: at one level involving capitalists themselves, each of whom strives to increase or maintain profitability, and at another between individual labourers seeking or merely retaining employment.⁵⁴ Labour market competition is characterized by him

---

52 Adherents of this positive view included not just Fanon (1963) and Worsley (1972, 1984), but also Hall (1960, 1989). The latter argued that, because of the co-optation of the industrial worker by metropolitan capitalism, class struggle in such contexts should be replaced with mobilization based on non-class identity ('cultural recovery') rooted in a non-/pre-capitalist Third World, thereby paving the way for the 'new' populist postmodernism which emerged subsequently.

53 Having traced the historical origin and emergence of the proletariat, Engels (1975: 375) then turned his attention to how capitalists sought to use labour market competition in order to block any advantage workers might derive from this development: 'Such are the various ways and means by which competition, as it reached its full manifestation and free development in modern industry, created and extended the proletariat. We shall now have to observe its influence on the working class already created [and] must begin by tracing the results of competition of single workers with one another'.

54 Competition, notes Engels (1975: 375), 'is the complete expression of the battle of all against all which rules in modern civil society. This battle, a battle for life, for existence, for everything ... is fought not between different classes of society only, but also between

as 'the worst side of the present state of things in its effect upon the worker', describing it as 'the sharpest weapon against the proletariat in the hands of the bourgeoisie'.[55] Thus it is precisely the most intense kinds of labour market competition that, argues Engels, gives rise to and generates in turn the most acute form of struggle between capital and its own workforce.[56]

Key to this struggle, and its deciding aspect, is how labour market competition undermines any worker solidarity based on class consciousness that might have existed hitherto. In the case of 1840s England, the attempt to roll back such gains as labour had managed to achieve, therefore, employers resorted to Irish migrants drawn from the industrial reserve, fuelling hostility that took the form of racism.[57] As Engels saw it, therefore, '[a]nother influence of great moment in forming the character of the English workers is the Irish immigration [which] degraded the English workers ... and aggravated the hardship of their lot'.[58] The threat this posed not just to hard-won wage levels and employment conditions but also to the protection of these gains – by means

---

the individual members of these classes ... the workers are in constant competition among themselves as [are] the members of the bourgeoisie among themselves'.

55 See Engels (1975: 376). Emphasizing how dependent capitalist production is on an 'unemployed reserve army of workers' he (Engels, 1975: 384) goes on to describe the kinds of economic activity undertaken by those belonging to the lumpenproletariat: 'This reserve army ... is the "surplus population" of England, which keeps body and soul together by begging, stealing, street sweeping, collecting manure, pushing hand-carts, driving donkeys, peddling, or performing occasional small jobs'.

56 'Hence the effort of the workers to nullify this competition by associations', observes Engels (1975: 376), 'hence the hatred of the bourgeoisie towards these associations, and its triumph in every defeat which befalls them'.

57 About the impact on class consciousness of this migration pattern, Marx (Marx and Engels, 1934: 289) noted: 'Owing to the constantly increasing concentration of farming, Ireland supplies its own surplus to the English labour market and thus forces down wages and lowers the moral and material position of the English working class. And most important of all: every industrial and commercial centre in England now possesses a working-class population *divided* into two *hostile* camps, English proletarians and Irish proletarians. The ordinary English worker hates the Irish worker as a competitor who lowers his standard of life ... [t]he Irishman pays him back with interest in his own coin. He regards the English worker as both sharing in the guilt for the English domination in Ireland and at the same time serving as its stupid tool. This antagonism is artificially kept alive and intensified by the press, the pulpit, the comic papers, in short by all the means at the disposal of the ruling classes' (original emphasis). Much the same point – about the divisive influence of immigration on political consciousness in the receiving context – is made by Engels (1975: 407), who observes that 'free competition rules, and as usual, the rich profit by it, and the poor ... who have not the knowledge needed to enable them to form a correct judgment, have the evil consequences to bear'.

58 See Engels (1975: 419).

of solidarity among and capacity of an existing workforce to organize – was such that Marx himself gave serious consideration to opposing further immigration.[59] On the subject of racism aimed by the existing workforce against 'other' immigrants, it is important to be clear – as was Engels – about what is meant by the ostensibly derogatory term 'uncivilized', applied to migrants from Ireland. As a result of class struggle and organization over many decades, local workers achieved a given standard of livelihood, wage level, and work conditions, which they seek to protect and build upon. This is not necessarily shared by the immigrant, who is – initially at least – prepared to work for less pay and worse conditions simply because the job, its remuneration and work arrangements are better than what is received or on offer at home.[60]

It is this economic dimension to which Engels refers when stating that 'one needs more than another, one is accustomed to more comfort than another'.[61] He is not talking about ethnic or cultural diversity (= supposedly innate/eternal racial characteristics) but rather about prevailing standards of living in different national contexts, a distinction based in turn on economic development, class formation, and the level of the productive forces, plus the respective

---

59 This was set out by Marx in a letter dated 9th April 1870 to Meyer and Vogt (Marx and Engels, 1934: 289–90), where to stem competition, he advocated severing the link with Ireland precisely in order to prevent migrants from competing with and undercutting English workers. Marx insisted that working class emancipation in England depended ultimately on Ireland following its own path of capitalist development, and to this end international solidarity would take the form of support from English workers for Irish equivalents in their struggle for economic and political independence, as distinct from migrating to where this had already occurred.

60 About this Engels (1975: 389) comments: 'The Irish had nothing to lose at home, and much to gain in England; and from the time when it became known in Ireland that the east side of St. George's channel offered steady work and good pay for strong arms, every year has brought armies of the Irish hither'. That the term 'civilization' as used by Engels applies not to ethnic or cultural 'backwardness' but refers simply to material conditions, those of economic deprivation, is evident from the next sentence, which reads as follows: 'These people having grown up almost without civilization, accustomed from youth to every sort of privation'.

61 This economic discrepancy, together with its implications, is charted by Engels (1975: 336–37) thus: 'Here we have the competition of the workers among themselves ... so that the bourgeoisie still thrives. To this competition of the workers there is but one limit; no worker will work for less than he needs to subsist ... True, this limit is relative; one needs more than another, one is accustomed to more comfort than another; the Englishman ... needs more than the Irishman ... But that does not hinder the Irishman's competing with the Englishman, and gradually forcing the rate of wages, and with it the Englishman's level of civilization, down to the Irishman's level'.

gains made in these separate locations as a result of class struggle over time.[62] This socio-economic difference does not prevent the migrant from competing in the same labour market as the local, much rather the opposite. Seen by the capitalist as a potential/actual worker, employment of the migrant possesses cost advantages the local does not have, or has no longer. The outcome is that, by entering the same labour market, the migrant reduces the wages and conditions enjoyed hitherto not just by the local but – eventually – by *all* workers. This is what Engels means when he talks of a reduction in the level of 'civilization' following an increase in migration.[63]

Conclusion

If one bypasses the overlooking or mistaken dismissal of all that went before, the global labour history approach to development espoused by van der Linden and Lucassen exhibits a worryingly conservative subtext. For his part, Lucassen seems to think that labour historiography can do without the concept of capitalism altogether. By contrast, van der Linden and cliometric historiography want to establish the presence in their respective contexts of capitalism, but each goes about this in different ways. Since van der Linden equates capitalism with unfree labour, the former is claimed to exist historically wherever the latter is found. Because cliometric historians equate capitalism with freedom, for the former to exist the latter must exist also. So chattel slavery is represented as a production relation akin to the free worker, a choice-making subject. What the analyses by van der Linden and Lucassen share, however, is a discourse that urges global labour historiography and/or social science generally to abandon Marxist theory once and for all. As such, it is an approach that invites – and should invite always – robust counter-critiques from those who remain Marxists.

---

[62] As noted widely at the time, the same kind of difficulty arose following the 2004 enlargement of the European Union, when less economically developed eastern European countries gained membership. A consequence of existing differences between lower living standards, wage levels, and work conditions in these countries when compared with equivalents in longstanding member states was largescale migration from the former to the latter. The increased labour market competition in the receiving nations resulted in a populist backlash.

[63] The blame for this lack of 'civilization', as Engels (1975: 411) makes clear, is not the workers themselves but the racist treatment to which they are subjected ('There is, therefore, no cause for surprise if the workers, treated as brutes, actually become such.').

This is because the problem of who is and who is not to be included conceptually as belonging to the proletariat has always been central to Marxist theory, which has taken a very clear and tough line on this issue. Clear because of its political importance, and tough because of the ever-present temptation to mobilize politically on the basis of an expanded inclusiveness of the term (= 'extended working class'). As long argued by Lenin, Kautsky, Trotsky, Engels, and Marx himself, there are numerous plebeian elements – an undifferentiated peasantry, lumpenproletarians, petty traders – whose economic and political interests are not just different from those of the working class but actually opposed to the latter. Any attempt to rope these elements into a broad anti-capitalist mobilization is bound to fail, as many studies about grassroots movements in the so-called Third World underline. This stems from the fact that such attempts do not distinguish between those standing resolutely against capitalism *per se* and those antagonistic only to certain kinds of accumulation (finance capital, foreign capital, agribusiness enterprises).

Hence capitalism is to be opposed – if at all – not by a progressive transcendence in the shape of a socialist transition, but rather by one of two kinds of regression: by a return either to a pre-capitalist rural idyll or to a 'nicer' version of the same systemic mode. Whichever of these two options is followed, the concept of modernity as a desirable/feasible project that historically has sustained development studies (especially in the three post-war decades before the rise of neoliberalism) appears now to have been sidelined, if not abandoned. These alternatives to a forwards-looking systemic transformation – as has long been argued – are those of populism and populists everywhere, a form of anti-capitalism which emerged from the political right, and has nothing to do with the progressive anti-capitalism of the political left. An inability to distinguish between the two forms is these days often encountered in the social sciences, and has serious political consequences.

CHAPTER 6

Liberalism and Development: Fukuyama's Scylla and Charybdis

> Nothing is so dangerous in a democracy as a safeguard which appears to be adequate but is really a façade.
>
> A dissenting note appended to a 1932 government report on ministers' powers by ELLEN WILKINSON, the radical socialist British Labour Party MP for Jarrow, warning against being taken in by reformist measures aimed at making capitalism palatable: As true now as it was then.[1]

∴

Introduction: a Benign Capitalism?

It is perhaps still important to remind ourselves of the issues raised by attempts to justify capitalist development, not least in the current ideological context where resurgent nationalisms and populisms are once again on the march. Now, more than ever, the question posed is which immediate development objective should leftists support: yet more nationalism, on the grounds that the main enemy is an external imperialism, and hence non-metropolitan capitalism is still a progressive force, or socialism? The reactionary and non-progressive aspects of nationalism in the recent past and currently are hard to disguise. Such characteristics underwrite attempts by wealthy regions (Lombardy, Catalonia) inside European countries to separate from the wider national context by claiming they are culturally/nationally other, and always have been. The same kind of argument was used by the apartheid South African state but in reverse form, separating wealthy/white areas from the Bantustans, designated culturally 'other' where impoverished blacks were confined.[2] Similarly, when in late nineteenth century Austria capitalists replaced

---

1 See UK Parliamentary Committee (1932:138).
2 Perversely, this separate development policy of the apartheid state was justified in self-serving terms as to the advantage of black communities, a way of protecting the latter by providing them with their own physical space where 'authentic' cultural traditions/practices and local institutions could flourish unhindered.

unionized German workers with cheaper Czech migrants, leftist parties advocated splitting worker institutions, organization, and politics along ethnic/national lines, thereby laying the ground for the emergence and consolidation of the far right in Austria and Germany.[3]

Historically, the difficulty with nation-centric development informing discourse about imperialism is the kind of political economic model which it trails in its wake: namely, that as capitalism is a system still largely confined to metropolitan contexts (US, Europe), the nation invoking anti-imperialism has yet to experience such a transition. Consequently, the next step is a political alliance composed of a national bourgeoisie ('progressive'), workers, and peasants, against foreign capital and for a bourgeois democratic stage in which a benign/non-foreign accumulation will establish itself in the nation concerned.[4] This is a very old argument, encountered most recently in the 1960s development decade, when economic growth in newly independent Third World countries required expropriating foreign owners of key economic resources (land, mines) on which then to build an authentically national accumulation project. It was, in short, the semi-feudal thesis, much criticized by Marxists then and since. Initially postponed, socialism eventually vanishes altogether from this political agenda.

Frequently underestimated or ignored is the contribution by academics to this nationalist upsurge.[5] Why we are still having this conversation is due in part to the shifts in the dominant theoretical paradigm about Third World development that accompanied the rise of neoliberalism. Mimicking the logic of capitalism, entry into academic posts of leftists from the 1960s onwards, and consequently Marxism as a topic of study, licensed a process of competition/recognition within universities that quickly became a plethora

---

3   See Whiteside (1962, 1975).
4   This sort of argument – criticized by me elsewhere (Brass 2018a, 2021b, 2022b) – can still be found, for example, in Patnaik and Patnaik (2017), for whom India has yet to become capitalist, and for whom the main enemy continues to be an external Britain. Claims about the continuation of British imperialism/colonialism where present-day India is concerned, together with the view that it is just such a relationship which is holding back its economic growth, are misplaced. Not only are large amounts of British industry now owned/controlled by Indian capital, therefore, but the extent of wealth generated within India by Indians themselves underlines the well-established nature of an indigenous capitalist class. Between 2020 and 2021 the number of Indian millionaires increased from 689,000 to 796,000, whilst those in the sub-continent with a net worth of US$100,000 numbered 17 million. See 'India wealth managers cast net beyond big cities', *Financial Times* (London), 22nd August 2023.
5   For details, see Brass (2018a, 2020, 2023a).

of reinterpretation.⁶ The latter entailed adding to Marxism concepts and theory that were non- or even anti-Marxist, leading inevitably to its dilution and depoliticization. Rather than the disempowerment of class, and its political resolution in the form of struggle for a revolutionary transition to socialism, the desirable objective quickly became empowerment or re-empowerment of non-class identities, to be achieved without necessarily transcending the capitalist system itself. This underlined the extent of ideological overlap between neoliberal economic theory and postmodern cultural turn (free markets, free choice of identity).

By the 1980s, therefore, the focus of the social sciences and the humanities more generally was changing dramatically; away from the materialist framework of Marxism, deemed inappropriate for an understanding of processes, issues, and populations outside Europe, and towards the 'new' populist postmodernism, the focus of which was on the culturally empowering nature of identity politics. The latter approach was – and is – strongly antagonistic towards Marxist political economy, dismissed by postmodernists along with its conceptual apparatus of socialism/materialism/class as just one more kind of Eurocentric/Enlightenment 'foundationalism'.⁷ Postmodern hostility expressed towards all things Marxist involves a twofold process: a denial of its historiography and conceptual apparatus is accompanied by an insistence on their replacement – epistemologically and politically – by a populist approach together with its privileging of peasant, ethnic, and national 'otherness'. Marxism is declared irredeemably tainted by a historical depriviliging of these same non-class identities that many postmodernists essentialize, in effect recuperating and proclaiming as empowering all the categories, processes, and identities criticized hitherto by leftist political economy.

Undermining worker solidarity, and thus also 'from below' organization based on class, populism has always been – and remains – one of the most effective 'from above' forms of struggle waged by capital. Where accumulation generates and feeds off an industrial reserve that is now global in scope, the

---

6   Among the reasons for this focus is that pointing to an enemy abroad is invariably safer than taking issue with one at home. Of additional importance, as indicated both elsewhere (Brass, 2017b: Chapter 18; Brass, 2018a) and in Chapter 4 of this volume, post-1960s entry into well-paid permanent academic posts, where membership of the Senior Common Room quickly deradicalizes any remaining leftist politics. Those who stay leftist are quickly shown the door, or never get through it in the first place.

7   See, for example, Escobar (1995), whose postmodern approach rejects development as an inappropriate foundational/Eurocentric model imposed by Marxists on rural populations of the Third World, a view shared both by the subaltern studies project associated with the work of Guha (1982-89) and by the 'everyday forms of resistance' framework of Scott (1985, 2012).

combination of disempowered Marxism and empowered populism is ominous. Just as some leftists in academia and elsewhere replaced core beliefs (socialism, class) with postmodern notions of non-class identity as innate/empowering, so the far right has in turn moved onto the vacated political ground, incorporating plebeian identity into its own ideology. To the postmodern argument emphasizing the cultural identity of the migrant-as-'other'-nationality, therefore, the far right counterposes an argument similarly privileging cultural identity, only this time the nationality of the non-migrant worker (= American, British selfhood).

Although the focus of this argument is on accumulation, and whether its continuing role in generating economic development is progressive, the issue is also and inevitably about something else: the political shape of the future. What it confirms is that a long-standing debate on the left, about the limits to nationalism, to bourgeois democracy, and ultimately to capitalism itself, is still relevant. This discussion, centrally, has been and is still on the issue of when, finally, Marxists can and should move to replace these combined systemic forms by putting socialism on the political agenda, and organizing/mobilizing to achieve this particular objective. If one takes socialism out of the equation, what remains is an opposition to capitalism that quickly arrives at the conservative position occupied historically by populists. The backwards-looking ideology of the latter combines an aggressive nationalism with the return to and restoration of a 'nicer' capitalism, not a transcendence of the system in question, and certainly nothing along the radical lines of a transition to socialism.

Accordingly, a long-term result of on the one hand supporting a development path linked to bourgeois democracy, accumulation, and nationalism, and on the other of demoting both the importance of struggle by workers to transcend capitalism and establish socialism, has been that in periods of economic crisis, when workers desire radical political solutions to their predicament, far-right populists colonize the political space that leftism has ceased to occupy. Insofar as it privileges cultural identity as empowering, current populism feeds off *laissez-faire* accumulation where economic crisis – generating both an expanding industrial reserve army of migrant labour and also more intense competition, between capitalists themselves and between workers seeking employment – results in political crisis. In the absence of a radical leftist discourse advocating a break with capitalism and a socialist transition, therefore, workers are encouraged by populists to experience labour market competition as an effect of non-class identity.

Currently and historically, evidence suggests that in these circumstances a radical politics remains on the agenda, but with the difference that working class support can be transferred instead to right-wing populist movements offering empowerment on the basis of nationalism and/or ethnicity. Where this eventually leads, no Marxist should need reminding. This is not the case, however, with non-Marxists, and especially so of academics who continue to defend capitalism on the grounds that it embodies liberal democratic norms. Perhaps the most prominent cheerleader of this approach is Francis Fukuyama, whose views about development merit dissecting. His latest book is significant in one way, and one way only: it shows not just how its defenders try to legitimize capitalism, but also how weak and unpersuasive the resulting case turns out to be. Because it is 'under severe threat around the world today', he announces that his object is to provide 'a defence of classical liberalism'.[8] The latter is equated by him with a seventeenth century doctrine emphasizing the necessity of protecting individual freedoms by placing constitutional or legal constraints on state power. As well as limited government. classical liberalism is a moral project based on law, freedom, and equal political rights, embodying values such as tolerance, progress, and liberty of the individual; the latter consequently possesses the right to engage in free economic activity and own private property without state interference.[9]

This defence rests centrally on the proposition that, despite these shortcomings, liberal democracy is in the end worth saving because it – and only it – is the political system that enables the masses to protect their interests *via* government, and thus accurately reflects any/all desires expressed by 'those below'. Just as identity politics is regarded by Fukuyama as a leftist cultural 'deviation' from authentic classical liberalism, so he is equally adamant that neoliberalism is a right-wing economic and political deviation from the same core doctrine. Each 'anomaly' is seen by him as an alien departure from true liberal values, and as such is to be condemned. Discontent stems not from liberalism itself, therefore, but rather only from the way in which 'sound liberal ideas' have been misapplied in ways unconnected with the doctrine itself.[10]

---

8   Fukuyama (2022: VII). Others engaged in the same kind of quest, a search for a way to protect and defend capitalism and bourgeois democracy, by rejecting neoliberalism and a return to what is claimed to be a more benign version of the accumulation process, include Stiglitz (2024).

9   According to Fukuyama (2022: 1–2), therefore, 'Liberal societies confer rights on individuals, the most fundamental of which is the right to autonomy, that is, the ability to make choices with regard to speech, association, belief, and ultimately political life. included within the sphere of autonomy is the right to own property and to undertake economic transactions'.

10 This process of epistemological distancing could not be more clear. 'Both are driven by discontents with liberalism that do not have to do with the essence of the doctrine', insists Fukuyama (2022: X–XI), 'but rather with the way in which certain sound liberal ideas have been interpreted and pushed to extremes'. This view is emphasized subsequently (Fukuyama, 2022: 17): 'Liberalism has seen its core principles pushed to extremes by

From this is drawn the conclusion that, as liberalism is essentially benign and still relevant, the object 'is not to abandon liberalism as such, but to moderate it'.

This chapter is composed of four sections, the first of which looks at what Fukuyama understands by classical liberalism, while the second and third examine how Fukuyama characterizes both neoliberalism and identity politics. His interpretation of liberal ideology about class, the industrial reserve, economic growth and agrarian reform is considered in the fourth section.

Floreat Classical Liberalism?

With the fall of the Berlin wall in 1989, and the subsequent disintegration of actually-existing socialism in the USSR and elsewhere, many academic commentators euphorically proclaimed that, as the Cold War had ended in the victory of capitalist democracy, the world would as a result become more peaceful.[11] Fukuyama was in the vanguard of those making this argument. Having announced in a 1989 essay to much public fanfare (and media acclaim) that history had come to an end, due to the defeat of communism by liberalism, democracy, and capitalism, he concluded that henceforth the absence of conflict between the socialist East and the capitalist West meant that there would be no more history to discuss. Such a view was then enshrined in a 1990s book, advancing the same triumphalist argument that, because the demise of the Soviet Union in his view consecrated the world-wide victory of liberal democracy, history-as-process had itself come to an end.[12] Events since then have, somewhat predictably, underlined the extent to which this prognosis was not merely over-optimistic, but much rather hugely mistaken.

Where Fukuyama is concerned, it is impossible not to admire his chutzpah.[13] His initial position was unambiguous: capitalism had won, and peace

---

        advocates on both its right and left wings, to the point where those principles themselves were undermined'.

11   Shortly after the break-up of the Soviet Union, Geoffrey Hawthorn, a sociologist in the Social and Political Sciences Faculty at Cambridge, made precisely this claim in the course of a seminar discussion about future developments in the global economic order.

12   For this view, see Fukuyama (1992).

13   Being so wrong, so often, has not dented the analytical and political confidence with which Fukuyama promotes the views he holds. Thus, for example, the following self-assured pronouncement (Fukuyama, 2022: XIII): 'I have spent a great deal of my life researching, teaching, and writing about public policy, and have no end of ideas about specific initiatives that could be undertaken to improve life in our contemporary liberal democracies'. This is reinforced by an observation towards the end of the book (Fukuyama, 2022: 147)

and happiness would now descend upon a world made safe for benign rule by capital, signalling that henceforth life across the globe would be a conflict-free one marked by sweetness and light. Having been spectacularly wrong-footed by all that has happened since, in almost everything written subsequently Fukuyama has attempted – unsuccessfully – to defend this original 1989 view whilst asserting the worth and desirability of liberal democracy.[14] Announcing that 'I do not intend this book to be a history of liberal thought', Fukuyama signals a wish to avoid engaging with internal contradictions, which leaves his analysis open to unfavourable comparison with other theoretical approaches to the same political ideology.[15] Hence the privileging by him of individualism as an epistemologically core element of classical liberalism, to the extent of acting as its cheerleader, is not a view shared by earlier writers on the subject who, although sympathetic to the political aims of doctrine, were nevertheless far more aware of its insurmountable political contradictions and deficiencies.[16]

Lacking any sense of contradiction, the case made by Fukuyama is littered with instances of unacknowledged or unresolved conflict, an aporia which requires that, as evidence of irreconcilability mounts up, the adoption of two contrasting arguments. On the one hand, ever more strident assertions on the lines that 'the basic doctrine is correct', while on the other more and more issues are decanted into a residual category labelled 'anomaly', 'shortcomings', 'extreme'/'untoward' departures and/or 'deviations' from the liberal norm.[17]

---

which reads like a pitch for a job as a government advisor: 'Modern states have to deal with a whole range of complex policy issues … and they need access to well-educated professionals with a strong sense of public purpose if they are to do their jobs well'.

14   See, for example Fukuyama (2018: XII–XIV), where he contends *The End of History* thesis referred simply to the case that 'a liberal state linked to a market economy was the more plausible outcome' of the development process.

15   Fukuyama (2022: XI).

16   Arguments deployed by Fukuyama in support of classical liberal doctrine do not survive a comparison with those made by, for example, Benedetto Croce, Michael Oakeshott, Ernest Barker, and John Dunn. Although none of the latter is by any stretch of the imagination politically leftist, each has analysed liberalism and made serious contributions to an understanding of its history, politics, and theory (Oakeshott, 1939, 1962: 37–58; Croce, 1941; Barker, 1951; Caponigri, 1955; Dunn, 1990, 2005, 2014). The extent of the difference between their view of liberalism and that of Fukuyama can be illustrated with reference to an earlier contention by Oakeshott (1939: XVII, XVIII) that 'the doctrine I wanted to represent is not to be confused with the crude and negative individualism which is apt to be associated with Liberalism … Liberalism in that sense is perhaps dead … Ignorant people are still to be found writing as if the history of Liberalism were merely the history of the rise and dominance of a peculiarly narrow brand of individualism; and if their ignorance were not dangerous it might be ignored'.

17   For these views, see Fukuyama (2022: XII, 12, 24, 31, 45, 76, 98, 150). Similar attempts to justify a line taken earlier inform his views about US intervention in Afghanistan and Iraq

This is compounded by a somewhat partial historical trajectory presented in support of the case being made. In the end the case on behalf of liberal doctrine amounts to nothing more than the hope that people should try to be nice to one another, a platitude which fails to indicate why in certain situations this is not possible.

Even in terms of the definition as envisaged by Fukuyama, the core elements informing classical liberalism admit of no contradiction, despite the obvious presence of the latter, not least because of a failure to see the systemic link between apparently disparate components/issues/processes. As a political doctrine, classical liberalism boils down to the autonomy of the individual to do more or less what s/he likes, both culturally (religious belief, practicing traditional customs) and economically (market transactions, property ownership), regardless of what the state might want to do in terms of policy. Thus a government in a Third World nation with a popular mandate to carry out an agrarian reform programme that entails land expropriation and its redistribution in different tenure forms (cooperatives, collective farms) would in effect constitute a negation of the very autonomy enjoyed by those whose existing property rights were affected.

I Am a Nice Shark …

Ironically, championing the classical principles of liberalism in this fashion cannot but end up on the same epistemological terrain as neoliberalism, protests to the contrary notwithstanding. Predictably, Fukuyama concludes by reasserting the primacy of individual rights, opposed by him to cultural rights (= postmodernism) not the collective rights based on class.[18] Endorsing the autonomy of the choice-making individual engaged in market transactions because in his view it cannot be changed, leads Fukuyama directly into the embrace of neoliberalism, which similarly justifies flexibility (= restructuring + cheap labour-power + industrial reserve army) and labour mobility (open-door immigration, yet more labour market competition) on the grounds that it is impossible to 'buck the market'. His dilemma is clear, and involves legerdemain akin to trying to explain the lifecycle of a shark without reference either to the sharpness of its teeth, its feeding habits, or the impact of its predatory nature on other creatures in the sea. The latter aspects are instead presented

---

(Fukuyama, 2006a, 2006b), presented either as 'nation-building' or as an unwarranted departure from the principles of neo-conservatism.

18 Fukuyama (2022: 150).

merely as 'anomalies', deviations from the true norm which, it is claimed, constitutes the essence of being-a-shark.[19]

Where economic ideology is concerned, therefore, Fukuyama accepts that deregulation and privatization informing the *laissez-faire* policies adopted throughout Western capitalist nations from the 1980s onwards meant that 'one of the critical domains in which liberal ideas were taken to extremes lay in economic thought, where liberalism evolved into what has been labelled "neoliberalism"'.[20] Despite the latter combining principles that are central to classical liberalism – the right to economic autonomy, limits placed on government – in his opinion the doctrine itself is unjustly blamed for 'the inequalities of capitalism'.[21] Such a contention – decoupling classical liberalism and its neoliberal variant – is counterposed by him to the opposite view: namely, that neoliberalism is neither an 'anomaly', an 'excrescence', nor a negation of doctrine itself, but much rather its authentic systemic realization, the economic actualization of the dynamic that is classical liberalism.[22] Against this, Fukuyama simply asserts – unpersuasively – that such critiques 'fail to hit the target, and amount to a charge of guilt by association'. That neoliberalism has fuelled the rise not just of oligarchs and tax havens but also of populism is for him a deviation from what he presents as the classical liberalism.[23] All these negative processes

---

19   Fukuyama's positive image of a liberalism decoupled from all the negative aspects of capitalism brings to mind nothing so much as the way the character of three sharks is depicted humorously in the animated Disney/Pixar film *Finding Nemo* (2003), directed by Andrew Stanton. Each shark maintains – implausibly – that it is now a reformed creature, a fish-friendly vegetarian species that has abandoned its predatory habits ('I am a nice shark, not a mindless eating machine').

20   Fukuyama (2022: 17, 19).

21   Fukuyama (2022: IX).

22   Hence the dismissal by Fukuyama (2022: 74) of the 'critique of liberalism [which] argued that the doctrine could not be dissociated from the most rapacious forms of capitalism, and therefore would continue to produce exploitation and gross inequalities, [a connection that] was not contingent but inevitable: liberalism with its emphasis on individualism and property rights inevitably leads to neoliberalism'.

23   See Fukuyama (2022: 29). The current role of neoliberalism in turbo-charging an already well-established economic inequality, not least the capacity of oligarchs to purchase government influence, is perhaps best described in a novel by Peter Green (1957) about ancient Rome. Opposed to the very idea of democracy, Sulla seized power in a 82/81 BC coup and ruled Rome as a dictator, establishing thereby the precedent followed not just by Julius Caesar but by all Emperors who ruled after the fall of the Republic. As depicted in the novel about Sulla (Green, 1957: 70, 123), this episode was itself both prefigured and licensed by the hollowed out nature that political representation had become in Republican Rome, a situation apparent to Jugurtha, King of Numidia, who looks at Rome and says: '"A city ripe for sale, and ripe for ruin, if only a purchaser can be found"'. It is a view repeated subsequently by Sulla himself: 'A city for sale, Jugurtha had said … if only a

or phenomena are seen as unconnected with what for him is the virtuous core of liberal doctrine.

Whether he likes it or not, what Fukuyama sanctifies as classical liberalism, a doctrine he wishes to recuperate, is based on the very principles that also underwrite *laissez-faire* ideology which – in the form of neoliberal economic theory – he insists is an 'anomaly', a 'deviation' from the essence of this same political doctrine. Unsurprisingly, therefore, he has difficulty in preventing his inner neocon from poking through the liberal carapace, the resulting paean to the virtues of neoliberalism in effect undermining the attempt to decouple it from classical liberalism. That Fukuyama still adheres to the view that neoliberalism is economically positive is evident from numerous examples, among them the following claims: '[t]he Reagan-Thatcher neoliberal revolution was grounded in, and solved, some real problems'; '[i]n the United States and other developed countries, deregulation and privatization had beneficial effects'; UK nationalized industries 'were more efficiently run by private operators'; the 'basic doctrine [of neoliberalism] is correct'; although there were 'problems with neoliberal policies ... this does not make it wrong'; and '[t]he problem with neoliberalism in economics was ... not that it began from false premises [since] its premises were often correct'.[24]

Both the fact and the closeness of a link between classical liberal doctrine and its neoliberal variant also emerges from a source not considered by Fukuyama: the French liberal school associated with the ideas of Turgot, Saint-Simon, and Dunoyer.[25] Formulated between the mid-eighteenth and mid-nineteenth century, their liberal political economy aimed *laissez-faire* theory at a landowning class which controlled both trade and labour by means of feudal legislation and monopoly over State power. It was to oppose this that the freedom of the population to sell their commodities – goods and labour-power alike – was endorsed. However, their struggle was confined to one against feudalism, so as to promote capitalism: it has underwritten *laissez-faire* economic

---

purchaser could be found', adding '"I know what your cheers are worth, to the last copper. I know how each section of the electorate was bought, and through whom. You are mine because I paid for you. Jugurtha was right."'

24 For these positive assessments of *laissez-faire* see Fukuyama (2022: 20–21, 22, 24, 31, 45). In keeping with these endorsements of neoliberalism are views (Fukuyama, 2022: 38, 92) that 'Hayek was right about the superior allocative efficiency of markets', and that neoclassical economics is simply 'a neutral application of the scientific method to the study of economics'. Unsurprisingly, Fukuyama (2022: 22) approves of the class war that Thatcher waged against organized labour, describing the conflict with the miners' union as her 'most heroic moment'.

25 For details, plus sources, see Brass (2018a: Chapter 4).

theory ever since, and is based on the view that capitalism is a positive good, an end in and of itself.²⁶

Much like Fukuyama himself, this branch of French liberal political economy maintained that in a 'pure' industrial capitalism a harmony of interests would prevail between capital and labour. Its object was the attainment of 'pure' industrial capitalism, to be realized by a peaceful transition to a *laissez-faire* economic system, in which producer and free labour compete in perfect market conditions. Once industrial capitalism in its most 'pure' systemic form – *laissez-faire*, in other words – is established, struggle ceases because the State, the principal locus of conflict and exploitation, is no longer under the control of feudal landowners. Significantly, this element of freedom was not confined to workers of a particular nationality, since the right of free sale of labour-power extended also to foreigners coming to or residing in the nation concerned. Established thereby was the principle underwriting the formation and operation – on a global scale, eventually – of the industrial reserve.

A Progressive Left?

Identity politics is viewed by Fukuyama as the second of the two main threats faced by classical liberalism, one that is cultural in nature and comes from the left of the political spectrum. The danger as he sees it concerns the erosion by 'left-wing progressive movements' of 'tolerance' for 'other life-style choices and values', which not only offends against a principle (autonomy) that is at the heart of classical liberalism but also fuels the backlash from the populist right that similarly constitutes a menace to the doctrine.²⁷ According to Fukuyama, therefore, when combined with deconstruction, '[t]he critical theories attached to identity politics in the United States [that extend] from structuralism through post-structuralism [to] postmodernism', were as a result 'carried to unsupportable extremes'.²⁸ Among those indicted for this 'leftist' postmodern deviation are Lacan, Barthes, Derrida, and Foucault.²⁹ Decoupling liberalism and identity politics, Fukuyama emphasizes that, contrary to the 'extreme' views held by these exponents of postmodernism, 'liberalism differed sharply

---

26 Marx departed from this, arguing that freedom was desirable because it prefigured the next step for workers *as a class*: namely, the capacity to organize politically so as to pursue a transition to socialism.
27 Fukuyama (2022: 17).
28 Fukuyama (2022: 85, 95).
29 Fukuyama (2022: 88, 89).

from nationalist or religiously based doctrines that explicitly limited rights to certain races, ethnicities, genders, confessions, castes or status groups'.[30] A consequence of endorsing identity politics, complains Fukuyama, is that 'many of the arguments pioneered by the progressive left have drifted over to the populist right'.

However, criticism both of postmodern identity politics and of 'religiously based doctrines' does not prevent him from endorsing the same core elements – patriotism, cultural tradition, religious belief – belonging to their discourses.[31] Conceding both that '[i]dentity politics makes liberalism difficult to implement', and that '[n]ational identity represent obvious dangers', Fukuyama laments the 'frequent inability' on the part of liberal societies 'to present a positive vision of national identity to their citizens'.[32] In keeping with his contradictory views about neoliberalism, his conclusion is 'not that identity politics is wrong, but that we must return to a liberal interpretation of its aims'.[33] Where identity politics is concerned, Fukuyama seems to understand neither its origin, its positioning on the left/right ideological spectrum, nor the nature of the dynamic governing its reproduction.

Disregarding the way the meaning of non-class identity is formulated, reproduced, why, and by whom, identity politics is reduced by Fukuyama to innate psychologistic aspects, an approach which overlooks the central role of capital in fostering ethnic/national differences among its workers, a long-established divide-and-rule tactic used by employers in the class struggle.[34] Ignored thereby is the complicity of liberalism in this process: insofar as capitalism is in ideological terms sponsored and validated by liberalism, the latter cannot be regarded as external to such a practice. This involvement in the capitalist project underlines the participation of classical liberalism in the rise and consolidation of identity politics, suggesting the latter is not quite the excrescence or 'anomaly' he would have us believe.

Fukuyama also misjudges the location of identity politics on the ideological spectrum, and like others adheres to the mistaken view that postmodernism is a leftist theory. Observing the support by some 'leftists' for the ideas of Carl

---

30 Fukuyama (2022: 2).
31 Hence the view (Fukuyama, 2022: 137) that 'Liberals have tended to shy away from appeals to patriotism and cultural tradition, but they should not. National identity as a liberal and open society is something of which liberals can be justly proud' In a similar vein, Fukuyama (2022: 146) maintains that the presence of 'deeply held religious beliefs' ought to be accepted.
32 Fukuyama (2022: 65, 129, 135).
33 Fukuyama (2022: 98).
34 Fukuyama (2022: 65).

Schmitt, the Catholic legal theorist who justified the seizure of power by the Nazis, Fukuyama misinterprets this as evidence that the left has turned rightwards, whereas postmodernism was never leftist in the first place.[35] Contrary to what he imagines, therefore, it was because they were not leftist that exponents of postmodern theory supported a rightist thinker: the reason that some postmodernists endorse Schmitt is precisely because they, too, are part of the populist right.[36]

As problematic are the origins of identity politics as seen by Fukuyama: affixed by him to the 1960s United States, when 'it got its start on the left' and attributed by him to Rawlsian 'justification for the liberation the inner self'.[37] This is incorrect. Within the United States itself, it was prefigured not only by the Ku Klux Klan (as Fukuyama accepts) but also by the plethora of hyphenated selfhoods (Irish-Americans, Italian-Americans, etc.) that constitute the 'melting-pot' approach to assimilation. Outside the United States earlier forms of identity politics have much deeper roots, going back at least to the Romantic reaction against the 1789 revolution in France, and the privileging of culture, nationalism, and tradition over notions of modernity/progress/development emerging from political economy. Many of the elements comprising the discourse of the 'new' populist postmodernism – among them the cultural turn, peasant essentialism, subaltern studies, everyday forms of resistance, multitudes, and proletarian multiverse – emerged from within development studies as a reaction against Marxist interpretations of economic growth in Third World countries. All of which suggests that, *pace* Fukuyama, identity politics is not merely not leftist but much rather anti-leftist.

Political Corrections, Problematic History

As noted above, the challenge Fukuyama makes against the case that liberalism and neoliberalism are inextricably linked – namely, that the latter is the inevitable result of the former – rests centrally on the claim that over the late nineteenth and twentieth century working class incomes rose while inequality declined, historical developments achieved only because 'advanced liberal

---

35 Fukuyama (2022: 82).
36 The overlap between postmodernism and conservative ideology is outlined elsewhere (Brass, 2000: Chapter 5; Brass, 2014b: Chapter 3), indicating how common to both is a desire to recuperate what are presented by each as historically innate, and thus enduring and non-transcendent, grassroots forms of national/cultural identity.
37 Fukuyama (2022: 47, 58, 62, 65–66).

societies' enacted social protections and labour rights. His twofold contention is: first, such progress demonstrates that capitalism and liberalism have in the past ensured 'political corrections' to any inequalities wrought by market economies; and second, '[t]here is no reason to think that such corrections cannot occur within a broadly liberal political framework in the future'.[38] In effect, Fukuyama ignores a whole range of objections to this weak defence of liberalism as a timelessly progressive doctrine.

To begin with, improvements in the position of the working class were a result not of liberalism but rather of class struggle conducted in opposition to capital and its state by organized labour, struggle which the liberal democratic state both opposed and outlawed. The argument that liberalism depends on democracy to regulate the market ignores the way in which the 'democratic process' actually functions: that is, how 'state capture' leads to policy and legislation which are frequently geared to the interests not of the working class but rather of those who are powerful and wealthy.[39] Similarly overlooked, therefore, is the Panglossian view that, when it is combined with democracy, liberalism can put in place 'political corrections' to offset the onward march of neoliberalism. This is a chimera, as is his conclusion that 'a broadly liberal framework' will in future effect such 'corrections'.

Because he has no notion of class, let alone of its formation, reproduction, and the fact of together with the reasons for the kind of struggles this generates, Fukuyama lacks also any notion of contradiction licensing a zero-sum process. Hence the realization of 'autonomy' by one individual or class is usually effected at the expense of the 'autonomy' enjoyed or desired by another. It is precisely because he has no sense of contradiction that Fukuyama is able to regard an all-embracing concept of 'autonomy' as a core element of liberal doctrine. Missing from his approach is the crucial process of struggle, its causes

---

38   This unwarranted idealism is also enshrined in the view (Fukuyama, 2022: 12) that '[l]iberalism's current travails are not new; the ideology has gone in and out of fashion over the centuries but has always returned because of its underlying strengths'.

39   The optimism on display at the outset, along the lines that in liberal societies the legal institutions which 'function semi-autonomously from the rest of the political system' (Fukuyama, 2022: 2) are neutral arbiters preventing 'state capture' by business, is seemingly reversed later. Hence the acknowledgement (Fukuyama, 2022: 29) that the ability of the rich to evade taxation by transferring money to tax-havens 'makes regulation very difficult' indicates the hollowness of democratic control. That money and power do indeed trump 'democracy' is an argument subsequently conceded by Fukuyama (2022: 92): 'Deregulation, strict defence of property rights, and privatization were pushed by wealthy corporations and individuals who created think tanks and hired big-name economists to write academic papers justifying policies that were in their private interests'.

and effects, involving not just capital and labour but also that occurring within the ranks of each. The latter takes the dual form: market competition between producers, and that between workers and migrants of different nationalities/ ethnicities, a rivalry for jobs played upon by employers.

Not denied is the fact that neoliberals support 'open immigration', and that restructuring whereby permanent/well-paid labour in metropolitan capitalist nations is displaced by cheaper foreign workers, either as a consequence of outsourcing the jobs themselves or insourcing migrant labour. Fukuyama admits that immigration results in job losses among those already in work, but nevertheless persists in his view that such restructuring is positive, in that it 'would lead to greater efficiency' and could be remedied simply by job retraining and welfare.[40] Ignored thereby is that in contexts where employers are able to draw on an industrial reserve army that is global in scope, as is now the case, intensified labour market competition is a zero-sum process: it means that liberal 'autonomy' is either that of the local worker (who retains his/her employment) or that of the immigrant (who supplants the local), but not both.[41] In

---

40   See Fukuyama (2022: 24–25, 37) who, like so many others, sees the difficulty posed by immigration mainly as a cultural one about citizenship/assimilation, and not principally as an economic one about the industrial reserve. Accordingly, any solution is positioned by him after the arrival of the migrant, rather than at the preceding stage – before citizenship/assimilation. This much is evident from his earlier observation (Fukuyama, 2018: 142–43) that '[t]he challenge facing liberal democracies in the face of immigration and growing diversity is … to define an inclusive national identity that fits the society's diverse reality, and to assimilate newcomers to that identity. What is at stake in this task is the preservation of liberal democracy itself'.

41   This process of capitalist restructuring, involving the displacement of well-paid labour with cheaper foreign equivalents, was illustrated somewhat dramatically during March 2022, when in defiance of legislative procedure P&O ferries went ahead and in a single day sacked 800 British crew, immediately replacing them with stand-by low-paid substitutes recruited from Eastern Europe, an act justified by the CEO in terms of the need to remain competitive with rival companies. See 'P&O halts ferry crossings and sacks 800 sailors via Zoom call', *Financial Times* (London), 18 March 2022; 'P&O sparks anger and chaos with mass sackings', *Financial Times* (London), 18 March 2022; 'P&O Ferries: keelhauled', *Financial Times* (London), 18 March 2022; 'P&O sets its course without fear of blowback', *Financial Times* (London), 22 March 2022; 'P&O's scant regard for employee protections', *Financial Times* (London), 22 March 2022; 'Staying Afloat: How P&O Ferries brought home ignored realities of a tough labour market', *The Guardian* (London), 30 March 2022. Significantly, this incident occurred just after the pandemic and Brexit was marked by a contraction in the industrial reserve, leading in turn to labour shortages in the transport industry, as a result of which existing workers were able to negotiate higher wages and better employment conditions. What this episode underlines is that as soon as the bargaining power of labour increases, so capital resorts to the industrial reserve in order to lower costs and maintain profitability. About this, democratic governments are either unable or else choose not to do anything.

these circumstances, plebeian concern about actual/potential job losses due to enhanced labour market competition quickly translates into political support for a populism undertaking to prohibit its source, 'open immigration'.[42]

In spite of the claim by Fukuyama that the 'durability [of liberalism] reflects the fact that it has practical, moral, and economic justifications that appeal to many people', it is clear that a valid role for liberalism as sponsor of 'economic growth and modernization' applies only to the beginning of the accumulation process.[43] Initially, therefore, the doctrine was mobilized in a progressive manner: the political struggle for accumulation, which entailed opposing feudal monopolies that posed obstacles to further development. Once capitalism had established itself as a system, however, the ideological target of liberalism changed: instead, the aim was now to resist a socialist transition, by preventing or frustrating political organization designed to transcend bourgeois democracy. In short, liberalism was no longer progressive, being interested in nothing but the survival of capitalism and its form of private property. So promotion of economic growth and modernization is historically-specific, not an intrinsic aspect of liberal doctrine.

As problematic in terms of history is the partial account by Fukuyama of the way agrarian societies were transformed in the post-1945 era, and the heroic role allocated by him to the United States in promoting land reform programmes throughout the Third World.[44] In Asia, therefore, where 'under American tutelage' large estates were broken up, 'redistribution of property has been widely credited as a basis for their subsequent economic success, not to speak of their ability to turn into successful liberal democracies'. This conveys the impression that the United States has been at the forefront of progressive land reform programmes (Korea, Taiwan, Japan) that led to economic growth and democracy, a triumph also for capitalism and liberal doctrine. Not mentioned, however, are instances – Guatemala in 1954, Chile in 1973 – where the United States helped oust democratically elected governments with a mandate to carry out agrarian reforms, ushering in instead anti-democratic

---

42   On the question of immigration, Fukuyama (2022: 127) is wrong to suppose that a 'leftist' society might simply operate an open-border approach, in effect allowing immigration to dilute citizenship, which would 'become essentially meaningless' as a result. Again, this does not take into account the Marxist position on the industrial reserve, which is not to promote but to oppose its enlargement; unchecked, the latter process enables capitalism to survive and prosper by turning migrants and locals against one another.
43   Fukuyama (2022: XII, 10–11).
44   Fukuyama (2022: 32–33).

dictatorships pledged to defend large estates by preventing such programmes of land redistribution from being carried out.

Conclusion

Just as the mythological beings Scylla and Charybdis were transformed into frightful monsters that preyed on passing seafarers, so liberalism turns into neoliberalism whilst its form of agency turns into identity politics, licencing a slide into populism. The latter sort of transformation, however, is one that is challenged and denied by Fukuyama. Legitimizing a hitherto questionable phenomenon frequently entails deflecting criticism by arguing that a more benign alternative exists, and can be supported politically. A 'solution' is found internally, one that does not require the transcendence of the phenomenon in question.

Hence the way Fukuyama seeks to justify the case for liberalism follows what is now an increasingly familiar pattern, both inside and outside academia. Examples of this procedure within the social sciences includes the defence of capitalism that contends it is an eternal (= non-transcendent) systemic form stretching from ancient society to the present. Pursuing the same objective with regard to liberalism, Fukuyama separates off what he claims is a core set of characteristics which currently are no longer adhered to, in effect splitting the concept between 'authentic'/positive and 'inauthentic'/negative variants, enabling claims that a return to the former is a solution to the problems inherent in the latter. In keeping with this are claims that as a 'nice' variant exists, consequently the solution to its 'nasty' variant (neoliberalism) is a return to this alternative version.

Similarly, in the case of populism its defenders claim that here, too, there are two variants, one 'nasty' (= authoritarian) and the other 'nice' (= progressive), and that the latter could – and should – be supported politically. Much the same kind of practice surfaces in the debates both about unfree labour and about the industrial reserve. By interpreting the latter as an issue to be addressed only or largely in terms of citizenship/assimilation, the focus of this approach remains on cultural 'values', not on the link between accumulation, enhanced labour market competition, and the industrial reserve. Those who maintain that unfree production relations can and will be eradicated by market forces make an analogous claim. Such positions shy away from questioning the accumulation process itself, which as a result is left intact. Just as these sorts of argument try to salvage capitalism by decoupling it from any and every negative systemic development (anomaly, deviation), so too does that of

Fukuyama, under the guise of recuperating an 'authentic' and 'benign' variant of liberal doctrine.

Fukuyama belongs to that category of academic which, no matter how egregiously wrong their prognoses turn out to be, nor how often such major errors of judgement are made, nevertheless manage to ascend virtually unhindered the career ladder in the western university system (senior posts, honorary degrees, institutional recognition, publications, media attention). It is a large category, one that extends from senior components of the academic structure, like Hobsbawm and Giddens, to a disparate agglomeration of more junior members lower down in the same hierarchy who compose the cheerleaders, providing support for those at the top. Unsurprisingly, an entirely predictable effect of this link between those occupying different positions in the hierarchy is a multiplier effect generating and reproducing egregiously implausible arguments. Difficult to avoid is the baleful conclusion that the institutional prominence of those in this category is in some degree connected with the ideological acceptability of their views to capital and its political representatives in the state apparatus.

CHAPTER 7

Anthropology and Development: Self in the World, World in the Self

> Now that I possess the secret, I could tell it in a hundred different and even contradictory ways. I don't know how to tell you this, but the secret is beautiful, and science, *our* science, seems mere frivolity to me now.
>
> Insights obtained by an anthropologist after fieldwork, as recounted in a short story by JORGE LUIS BORGES.[1]

∴

Introduction: What Do I Know?

Among the tropes that circulate perennially are questions about what sort of person writes an autobiography, and why. The purposes of such narratives are many, and extend from the simple desire for self-aggrandisement to handing down to subsequent generations an account of lessons learned and achievements garnered, the hope being that the ideas and/or agency of the author can be said to have 'made a difference'. In any list of those who do write such accounts, politicians and 'celebrities' feature prominently, their narratives hiding rather more than they reveal about the self.[2] For the most part, therefore, political memoirs, like those 'written' by 'celebrities', are ephemeral and insignificant documents soon consigned to the dustbin of history. In cases of autobiography written by a journeyman academic, the situation is somewhat different.

Not many academics write autobiographies, one reason being that few are deemed – by the public and publishers alike – to have anything of significance to say about their own lives or to have done anything that is noteworthy,

---

1 Borges (1998: 335, original emphasis).
2 There are exceptions, of course. In the UK a number of published diaries kept by politicians from different ends of the ideological spectrum – Richard Crossman (1979), Tony Benn (2007), and Alan Clark (2010) – are justly regarded as classics of the genre.

since what there is of interest is usually poured into their published work.[3] Intellectuals who have a world presence and impact, such as Keynes or Marx, leave it to others to examine and define the nature and motivations of their own selfhood, in the shape of biography. Rather than a full-length autobiography, for many senior academics the preferred method of communicating aspects of the self consists of an extended interview.[4]

To some extent, the popularity of latter form has been fuelled by the rise of postmodern theory that privileges subjectivity, especially that of an author who is an academic, now required to state who s/he is and why s/he writes. Erasing the boundary between historiography (as non-fiction) and literature (as fiction), this is the kind of approach that unhelpfully encourages a slide into aporia and solipsism.[5] Whereas the focus of much autobiography is on the world as seen by the self, a rare methodological advantage enjoyed by social anthropologists consists of an ability to draw on life histories of the 'other' constructed during fieldwork, so much so that it can reinforce any discourse featuring a combination of self-in-the-world and world-in-the-self.

There is accordingly a discernible pattern to autobiography written by academics, a template which usually highlights one or more of the following aspects. First, the presence of ancestors – internal (kinsfolk) and external (influences, teachers) – who are at least interesting, preferably famous, and with achievements to their name. Second, colleagues/friends who are also well-known, holding senior posts in world-famous universities, not just having achievements to their name too but also recognizing one's true worth. Third, a list of personal achievement, consisting of important discoveries – a theory, a concept – made before anyone else, a contribution that becomes central to

---

3   This contrasts with the way in which academics have been portrayed on film, ranging from *Pimpernel Smith* (1941), directed by Leslie Howard, to the *Indiana Jones* series (1981–2008) directed by Steven Spielberg, in which the protagonist – in each case an archaeologist – is invariably depicted as academic-as-action-hero engaged in fighting oppression and rescuing the downtrodden. The latter image, however, does apply in one particular case, that of Doreen Warriner. A distinguished academic who conducted research into rural development and wrote extensively about peasant economy in Eastern Europe and elsewhere (Warriner, 1939, 1948, 1950, 1957, 1969), she was also instrumental in rescuing Jewish and other refugees threatened by the rise of fascism in pre-war Czechoslovakia (Warriner, 2021). Where she is concerned, therefore, the kind of academic depicted in films by Leslie Howard and Steven Spielberg holds true.
4   See, for example, Lévi-Strauss (Lévi-Strauss and Eribon, 1991), Giddens (Giddens and Pierson, 1998), and Hobsbawm (2000). The latter also wrote an autobiography (Hobsbawm, 2002).
5   It could be argued that to the extent that they are discouraged by postmodernism from addressing issues concerning identities which they themselves do not share, social scientists are in a sense left with no alternative except to write largely about themselves.

the discipline in question. And fourth, a narrative that privileges adversity, whereby achievement is the outcome of personal struggle conducted against intellectual fashion that prevails within academia, if possible allied to victimhood as a result of persecution by authority or the state itself.

Some of these aspects can be found in the latest book by Keith Hart, an influential but unusual anthropologist, who at the age of 80 has written an atypical but intriguing autobiography which deserves to stand alongside the best of academic memoirs composed by social anthropologists or sociologists.[6] The reasons for this are fourfold. First, its form: he writes very well, a lesson in this regard for other academics. Second, its content: ideas/arguments are clearly presented, if not always consistent politically, and a life is recorded in a revealingly confessional mode. Third, its role as a form of *testimonio*: to some degree, the story told is one about the precariousness of the modern academic career, especially as this concerns the social sciences. And fourth, he has made important contributions to the study of development. Moreover, the expansive title is apt, since as Hart himself emphasizes, introspection ('This book is about learning and aspiration') extends outwards from the self, its purpose being also 'to find provisional answers [to the question "what does it take to become fully human"] by examining my own experience and reflecting on what I have learned of the world'.[7]

This chapter consists of two sections, the first of which looks at the narrative about internalities, or how the self has experienced the world. The focus of the second is on externalities, or how the world appears to and is understood by the self.

I

In many ways, the most interesting parts of the narrative concern details about the trajectory of Hart's life, in particular how he has managed to navigate health and academic employment problems by means of gambling and financial acumen, skills that also served him well in his anthropological fieldwork. This kind of information rarely makes it into the public domain, let alone ethnographies based on fieldwork, which tend to avoid mention of personal

---

6   Among them the ones by Elwin (1964), Powdermaker (1966), Malinowski (1967), Murphy (1987), von Fürer-Haimendorf (1990), Worsley (2008), and Eribon (2018). Unlike most of the latter, the focus of which is on encounters with the ethnographic 'other', Hart places as much emphasis on the self-in-the-world.

7   See Hart (2022: XI, 2, 3), who describes himself in the following manner: 'I am an anthropologist by profession, an amateur economist by inclination. I devour movies, novels, sport and all kinds of music. Most anthropologists discover the world by finding out what people do and think where they live. I rely mainly on lifelong learning through reading, writing and varied world experience that includes eclectic immersion in high and low culture'.

travails accompanying (and informing) the research process.[8] It is this as much as anything else that underlines the value of his autobiography.

The Self (in the World)

The main kind of adversity recounted by Hart involves ill-health, which has affected both his employment and writing. Blind in one eye since birth, he records ten mental breakdowns over a fifteen year period, and two in one year which to him 'suggested chronic insanity' meaning that he 'might never work again'.[9] Such events shaped in part the work he undertook: having obtained an academic post at Chicago, therefore, Hart observes that 'I had another mental breakdown as soon as I got there and that ruled out a permanent job'.[10] Suffering from the mood swings indicative of bipolar disorder, he notes that 'colleagues reacted to these episodes with kindness'. The 1983–86 period is described by Hart as his 'lowest ebb', when a capacity to write was obstructed by this medical condition, and it was only when he remarried in 2001 and had a settled family and children that, finally, he was 'freed from mental illness'.[11] Even so, the full extent and deleterious effect of health problems is difficult to avoid: 'I had been diagnosed with prostate cancer, a stroke, kidney failure, Parkinson's and neurological disfunction, not to mention psoriasis, mental breakdown and lithium poisoning … [m]y immune system, not to mention my confidence and will to live, collapsed'.[12]

If ill-health can be seen as a form of adversity straightforwardly, then the narrative about employment is less so. On the one hand, therefore, are experiences that are undeniably negative and personally disempowering. Thus, for example, when the social anthropology department at Cambridge advertised a junior post, Hart applied and was successful; now aged forty, however, he admits that he 'could not swallow the demotion of being on the bottom rung'.[13] After holding various temporary appointments in the United States, Hart accepts that 'I now found myself alone, overstretched and without prospects', as a result of which 'I would go to Cambridge without a job and see what

---

8   His account, Hart (2022: 17) notes, has 'included unflattering episodes'.
9   Hart (2022: 69, 110, 161).
10   Hart (2022: 114).
11   Hart (2022: 118–119, 142).
12   Hart (2022: 163–64).
13   For this and what follows, see Hart (2022: 118–119, 177).

turned up'.[14] Significantly, the numerous academic posts and consultancies held by Hart over the years follows a pattern that, although familiar now, was not so then.[15] His peripatetic experiences, an early instance of the academic as world subject, looking for and finding employment – of mostly temporary duration – in many different universities across the globe, prefigure what is currently more-or-less the norm.[16]

On the other hand, however, are instances of empowering participation in academia.[17] During the 1970s Hart found employment at the universities of Manchester and East Anglia, where he undertook consultancies on development issues, advising the governments of Papua New Guinea and Hong Kong.[18] Receiving job offers from Chicago and Yale, he accepted a tenured position at the latter, but then resigned in favour of a temporary appointment at Michigan, moving on to another post, this time at Chicago.[19] Since '[b]eing an assistant lecturer in Cambridge had already worn thin for me', Hart departed for Jamaica, obtaining a post at the University of the West Indies where he helped 'set up a graduate school for the social sciences in the English-speaking Caribbean'.[20] Subsequently, he was offered employment in the Washington University anthropology department, where Hart 'received the best job offer I ever had' – a full professorship with a high salary, but turned this down to

---

[14] In terms of the peripatetic nature of academic employment, combined nevertheless with a wide-ranging influence exercised on important debates in the social sciences, the trajectory and experience of Hart can be compared to that both of Tom Nairn and of Daniel and Alice Thorner.

[15] Emphasizing this difference, Hart (2022: 224) comments: 'In 1969, when I got my PhD, there were twenty-three lecturing jobs available. One had no applicants. The new universities were still recruiting, and their students had not yet reached the job market. This situation soon turned to one of job scarcity'.

[16] Having worked in twenty-four countries across four continents, 'a story of [a] nomadic life', Hart (2022: 6, 143–44) elaborates on the effect: 'I have worked part-time in a dozen universities. I have given public lectures in two dozen countries. I spend a of time online ... I embraced social media with enthusiasm after 2000 ... I planned to be a freelance writer ... I left everything I had in Cambridge. I had no job, institution or circle of friends ... The older I got, the more introspective I became'.

[17] The list of the-great-and-the-good in the academic world with whom Hart (2022: XI, 78, 79, 96, 107, 108, 119, 155) has rubbed shoulders is long and impressive: Audrey Richards, Edmund Leach, A.J. Ayer, Jack Goody, Edward Shils, Meyer Fortes, Marshall Sahlins, Frederic Jameson, David Apter, Umberto Eco, Catherine McKinnon, Ernest Gellner, and Stanley Tambiah. Much like Galbraith (1999), however, the description by Hart of these encounters is done in a non-boastful and matter-of-fact way.

[18] Hart (2022: 102ff.).

[19] Hart (2022: 110ff.).

[20] Hart (2022: 120).

work with C.L.R. James.[21] Returning to Cambridge in 1988 as a tenured lecturer in the social anthropology department, a Fellow of Girton College and Director of the African Studies Centre, a decade later he finally retired as a British academic, and left to live and work in Paris and Durban.[22]

Self-Help

In light of this chequered employment history, it is indeed fortunate that Hart possesses other strings to his bow, in the shape of an ability to generate income from gambling.[23] From an early age, therefore, he made money both as a result of becoming 'an expert card player' and from betting on horses.[24] This success extended from his undergraduate days onwards: whilst at university he 'won enough at cards to pay my share of party expenses', and thereafter '[f]rom cards and betting, I roughly doubled my income'.[25] On entering the academic job market, this income source became crucial: he 'had a $25,000 gambling fund

---

21 Hart (2022: 122). Observing that 'James deserves to be better known by anthropologists', Hart (1990a: 45) gives the following as reasons: 'His work occupied the imaginative space between local particulars and the broadest vistas of world history', adding that his 'hatred of academic specialisation and the consequent fragmentation of intellectual understanding led him to invent a model of what the best anthropology has always aspired to be … [his] own life – inside, outside and between the social polarities of our world – is an inspiration for what some of us hope anthropology may yet become'. Each of these reasons – epistemological unity of self and the world, dislike of anthropology as an academic discipline, and its application beyond the university – reflects the intellectual approach of Hart himself.

22 See Hart (2022: 129ff., 136, 147), who persuaded UNITA party members to attend a conference on Angola, one of the many events for which he 'recruited professionals, activists, scholars, and students'. Despite leaving the UK, he (Hart, 2022: 153) continued to hold various temporary and/or part-time academic posts in British universities: a full-time research fellowship at Aberdeen, a half-time professorship at Goldsmiths; and a 30% professorship at the London School of Economics. It remains unclear whether these appointments were chosen as such by Hart, or whether they were only available on these terms. In short, if they constituted an empowering or disempowering engagement with the academic jobs market.

23 Also mentioned by him (Hart, 2022: 64) is an inheritance of £100,000 received from an aunt.

24 See Hart (2022: 65, 65, 71, 76), where he recounts that 'after three years I was making a regular profit [gambling on horse racing, and] whenever I earned wages, my betting expanded'.

25 'I now had capital in Cambridge', he observes (Hart, 2022: 80–81): 'Six years spent grubbing around the bottom end of the betting market paid off. I knew a lot about horses … I made an average 8 per cent on turnover … I couldn't lose'.

in case of unemployment [and he] used it now to trade in dollar/deutschmark exchange rate futures'.[26] Later, when faced with a distinct lack of employment prospects, Hart reports that his 'betting stake had grown to $30,000 thanks to money derivatives'.[27]

As Hart makes clear, 'I learnt more about the world economy from playing the markets than from studying': success at gambling had a practical benefit, not only as an economic safety net but also in terms of anthropological research in Accra.[28] In the course of conducting fieldwork in a west African slum, the role he undertook was 'principally a receiver of stolen goods, then a moneylender'.[29] Among the participant/observation activity undertaken during this period, therefore, Hart 'forged receipts for stolen goods and fenced [items] seized by the police', things which he accepts were not possible to acknowledge in published or spoken form at the time: 'how hard it was', he comments, 'in those pre-postmodern times to make an academic career out of this material without lying'.[30] Consequently, Hart was unable truthfully to present his fieldwork findings: 'I had to convert all my stories into the third person. I gave a talk in Chicago in 1983, where I revealed some of it, and was told to shut the hell up or I would compromise Anthropology Inc. as well as myself'.[31]

This helps explain, perhaps, why Hart perceives in negative terms both academia in general and in particular social anthropology as an academic discipline.[32] What he opposed was that 'intellectual life [at Chicago] was cutthroat, with graduate students competing as proxies for rival professors', while Cambridge 'produced confident ignoramuses with fast mouths'.[33] Unlike others, however, he also took issue with the way 'young scholars perform cheap labour without prospects', a situation the more senior academics use to their own benefit.[34] At the same time, he views the pursuit of an academic career

---

26   Hart (2022: 114).
27   Hart (2022: 117).
28   Hart (2022: 165).
29   Hart (2022: 83).
30   Hart (2022: 83, 88).
31   Hart (2022: 94).
32   Since 'anthropology privileges a passive and narrow localism that has lost the nineteenth century's aspiration to understand human history as a whole', the laudable aim of Hart (2022: 29) 'is to revive modern anthropology's origin in the democratic revolutions of the eighteenth century, to show that there was a progressive anthropology before colonial empire and bureaucratic capitalism', adding for good measure that (Hart, 2022: 95) 'some of my academic colleagues are the most parochial people I know'.
33   Hart (2022: 107, 108).
34   'What was left of academic community was destroyed by the growing gap between a few established professors who took leave often and a reserve army of precarious young

on the part of these same young scholars as an obstacle to the formation of a popular anthropology outside the university system.[35] To this is attributed the failure of the Open Anthropology Cooperative to survive and prosper, its membership being too attached to the way it was defined and reproduced within academic institutions.[36]

No Friends There

Linked to this disenchantment with social anthropology as an academic discipline is its methodology: participant/observation fieldwork, a form of research which Hart associates with colonial attitudes to 'primitive' societies. Commenting that in anthropology '[t]he only knowledge rule was "I have been there, and you haven't"', his disillusion stems from the fieldwork experience itself, a point he emphasizes thus: 'My time in Ghana was an eye-opener ... the label "anthropologist" had no friends there'.[37] 'The anti-colonial revolution put paid to anthropology as the study of "primitive peoples"', remarks Hart, 'but most anthropologists have since clung to the narrow localism and ahistorical vision of "fieldwork-based ethnography" [and have] struggled to catch up with global events they did not understand nor could shape'.[38] Ironically, given

---

teachers', observes Hart (2022: 185), a development culminating in what is currently a highly exploitative division of labour (Hart, 2022: 226): 'Professors enjoy enhanced rewards and freedom. Young scholars perform cheap labour without prospects ... we now take leave for research and writing, while the university hires a young temporary replacement for a pittance'.

35 In what appears to be a form of self-criticism, realizing perhaps that he is himself guilty of the same career pursuit that he now finds problematic in young scholars, Hart (2022: 115) notes that 'I had long contemplated leaving academia. I could be a journalist, businessman, politician', but nevertheless opted for what he regards as 'my academic vocation'.

36 On this point see Hart (2022: 4–5), who remarks that 'we formed the Open Anthropology Cooperative [in 2009], which acquired over twenty thousand members from around the world' the focus of which was on 'how anthropology might evolve ... by becoming a popular rather than an academic pursuit'. Among its objectives (Hart, 2022: 181ff.) was the formation of a network of anthropologists who would contribute to an online discussion forum, and to enable young researchers to access journals and other publications hidden behind paywalls. However, the problem was that, despite wanting 'to avoid the academy's formality ... most members worked in universities'.

37 Hart (2022: 47, 80).

38 See Hart (2022: 4). Anthropology, which in the past involved mainly the study of small inaccessible units in faraway contexts, is for such practitioners now based more on comparative analysis over time (history) and space (geography). In terms of academic employment and fieldwork, therefore, there has been to some extent a fusion between anthropologist-as-global-employee and anthropology-as-global-research. As Hart

his opposition to fieldwork, he nevertheless accepts that 'the time in Ghana shaped me more than any other and still does half a century later'.[39] Hence the centrality to his ideas about development of the informal economy.

It is possible to surmise that the negative view Hart has of academia is also due to his engagement with this institution. Of particular interest, therefore, is the light he is able to throw on the origin of and debate about the 'informal sector' concept. Hart records how the term emerged as a challenge to 1960s modernization theory, a paradigm contending that developing societies in the Third World would follow the route out of poverty charted by metropolitan capitalist nations, which one day they would come to resemble. At this conjuncture, development theory was engaged in an attempt to understand the logic of urban exclusion in order to remedy this by means of modernization and so avoid it becoming a source of leftist political mobilization. The fear on the part of planners and politicians was that unemployed/underemployed rural migrants in urban slums formed an impoverished reservoir of antagonism/despair that might be harnessed by leftist movements, as such constituting a potential challenge to the capitalist system.

Against this a number of anthropologists who were conducting fieldwork in urban contexts, among them Hart in Ghana, drew a very different picture ('This story didn't square with my fieldwork experience').[40] Thus, for example, Anthony and Elizabeth Leeds challenged the innate fatalism of the 'culture of poverty' thesis formulated by Oscar Lewis, which depicted the urban slum as the locus of disempowerment because of its economic and social marginality, stressing much rather the economic and social dynamism of the *favela*, and redefining its inhabitants as empowered.[41] For his part, Hart argued that, far

---

(Grimshaw & Hart, 1994: 229–230) frames the problem, anthropologists 'have always derived their intellectual authority from direct experience of social life. In consequence, their "expert" knowledge is essentially commonplace, what everyone experiences, albeit in different forms, as a member of human society ... in general ethnographers traded in common sense, relying on the unchallengeable monopoly afforded by fieldwork in foreign places'. However, '[t]he accelerated integration of world society since the Second World War has severely embarrassed this project'.

39   Hart (2022: 95).
40   See Hart (2022: 99). 'I couldn't help noticing', he states (Hart, 2022: 86), 'the vitality of street commerce, not just roadside vendors, but mobile dealers in everything from refrigerators to Marijuana. I found that I was being drawn into local society through a variety of economic transactions – exchanging currency, shopping, paying rent and wages, making gifts, loans and bribes – all of which challenged my assumptions concerning what is normal. I embarked on the road to what became eventually the informal economy'.
41   See Leeds & Leeds (1970) and Leeds (1971). The latter – like Hart – contested the application by modernisation theory of negative concepts like 'marginality' to slum dwellers, arguing that – drawn from Western political economy – they misinterpreted or ignored

from being a locus of the uniformly downtrodden as depicted in 'the unthinking transfer of western categories to the economic and social structures of African cities', where the denizens of these seemingly peripheral areas were concerned, 'their informal economic activities possess some autonomous capacity for generating growth in the incomes of the urban (and rural) poor'.[42] The latter was a process to which Hart did indeed give a name, one that has stuck ever since and is nowadays used in many different ways.[43]

What happened in the course of subsequent debate about the informal sector is all too familiar. When in 1971 Hart presented a paper to a conference in the Institute of Development Studies (IDS) at Sussex University, about his fieldwork and the concept linked to this, two well-known economists responded negatively (they 'were sceptical'). Twelve months later, however, the publication of an ILO report about Kenya, jointly edited by these same two economists, centrally featured the informal sector approach, but now without any reference to the work of Hart, an absence which 'caused a minor scandal'.[44] He is aggrieved about this, rightly so: it illustrates a practice not wholly unknown in academic circles, about how new ideas/concepts formulated on the margins

---

the reality on the ground. 'We wish to make ... methodological and theoretical points' concerning the term 'integration', commented Leeds and Leeds (1970: 230, original emphasis), since 'it is currently being used in a vast literature on developing societies in a value-determined manner to mean "our", i.e. an American 'Western', 'democratic', 'price-making market', kind of 'integration'. Instead, '[i]t becomes plain that we must begin to think of qualitatively different forms of integration ... Social scientists should direct intensive effort at understanding these integrational forms *in their own terms*, as viable forms independent of the American or "Western" models'. For his part, Hart (2022: 97, 99) points out that 'I played a major part in the discovery of the "informal sector"', but accepts that it 'was only partially true' that he alone was responsible for the idea.

42 These quotes are from Hart (1973), where he presents the initial analysis of his Accra fieldwork.

43 As Hart (2022: 100) puts it, 'I became the contested founder of a major concept in development studies – 'the informal sector', later known as 'the informal economy' and eventually as 'informality', adding: 'I had no ambition to coin a concept. I just wanted to insert a vision of irregular economic activity into development debates'.

44 The two economists concerned, both IDS Fellows, are mentioned on page XIII of the Preface, while the informal sector itself is referenced throughout the Kenya report (International Labour Office, 1972: 5–6, 21–22, 51, 53–55, 68, 69, 76, 77, 119, 203, 223–32, 326, 341–43, 503–505, 506–508, 587–88). A comparison between the way the informal sector is defined, and of what it consists, both in the ILO volume and in the article by Hart, leaves no doubt, certainly in my mind, as to the extent of the influence exercised by the latter account over the former. Elsewhere Hart (1992: 216) recounts how one of the economists had initially 'arrived by dubious means at a figure of 30 per cent for urban unemployment', before going on to incorporate the informal sector concept within his own analysis.

of academia are taken over by those more senior, often after having dismissed them earlier.[45]

II

In a fundamental sense, the issues framing the way Hart sees his own experiences (self-in-the-world) also inform his interpretation of the world itself. Although this is unsurprising, the narrative about the life makes it possible to see this connection more clearly than is usual. Hence the method (gambling, playing the markets) in which he managed to generate additional funds to supplement the part-time/temporary nature of academic employment, finds a parallel in the self-sustaining economic activity encountered in his analysis of the informal sector. Each case – his own and that of slum dwellers – is sufficiently central so as to defy the label of marginality. That the anthropological self and the 'other' of anthropological study fuse in this manner, both adopting similar kinds of approach to the same issues, underwrites the humanistic ideology based on individual autonomy to which Hart subscribes.

The World (in the Self)

It comes as no surprise, therefore, that many of the views held by Hart about the nature of the world reflect his own experiences therein. It should be stressed that there is nothing untoward about connecting such perception to the way he depicts the self: what is unusual is that, when referring to intellectuals from whom he takes his inspiration, Hart – unlike others in the social sciences – makes both the link and the reason for these influences explicit.[46] Edward Gibbon and Giambattista Vico are commended for, respectively, encouraging 'self-learning' and being a 'teacher of himself', while Henry Adams 'got his education from … friendship and reading', and Nabokov 'is one of the great prose stylists', all much like Hart himself.[47] Similarly praised are three anthropological ancestors: W.H.R. Rivers, not only because his 'method became more self-reflexive' but also for his 'integrated vision of self and society'; and Marcel Mauss who, together with Emile Durkheim, 'wanted to make the social market visible'.[48]

---

45   For similar cases, see Brass (2014a; 2017b: Chapter 16).
46   More often than not, the political background to what is claimed to be a 'new' paradigm in the social sciences remains hidden. For one example of this, together with the epistemological problems generated as a result, see Brass (2023a).
47   Hart (2022: 20, 25, 26, 30).
48   Hart (2022: 39–43).

As well as Rousseau, Kant, and Marx, other major influences – all of whom are regarded by Hart as 'my guides to making a better world' – include not just Fanon, C.L.R. James, and Du Bois, but also Gandhi, whose 'ideal of bottom-up organization lives on through work organizations, voluntary associations and NGOs'.[49] This eclectic combination of intellectual ancestors from whom Hart draws inspiration points to a rather obvious difficulty: given their political incompatibility, what – if anything – do they have in common that could be said to underwrite an ideologically coherent worldview informed by a single theoretical approach? His answer to this question is humanism, a quasi-religious philosophy Hart equates with the anti-colonial intellectual, observing that '[t]he Pan-Africans combined religious thinking and secular politics … [t]he idea of soul was central for Gandhi and at first for Du Bois … Gandhi's religious politics were humanist'.[50]

Departing from the negative way Marxism defines religion – as a form of false consciousness – Hart seemingly regards such belief as positive.[51] This in turn can be traced back to when at an early age he subscribed to religious faith, associated by him (74) with humanism: 'I embraced the idea that we should love all humanity [and] recognize the human being in those we meet'. According to Hart, therefore, '[w]e must develop our self-knowledge as individuals and as a species. The relationship between the two is important [since] what interests me is how each of us relates to the whole', adding: 'How social divisions mediate that relationship is secondary'.[52] This is central to his argument about humanism; the latter is a common identity we all share, and thus overrides any other identity, including that of class and its focus on 'social division'.[53] Following Kant, Hart espouses the view that '[h]umanity's last task is the equal administration of justice everywhere', without, however, indicating how this concept of justice addresses – let alone reconciles – the antinomies inherent in social division.[54]

---

49   Hart (2022: 54ff.).
50   Hart (2022: 55–56).
51   'Traditional religion helps devotees to make a meaningful connection between self and the world', argues Hart (2022: 56), but as 'the rhetoric and practice of science have replaced the humanities and religion as a guide [so] we need to synthesize these poles somehow'.
52   Hart (2022: 37).
53   To this end, in 2011 Hart (2022: 148) established a Human Economy Programme at Pretoria University.
54   Hart (2022: 38).

Insufficiency

On the subject of Marxism, Hart displays conflicting views, oscillating between approval and disapproval.[55] At times, therefore, he appears supportive: hence references throughout the book to this effect, including the fact that he 'took up socialism as a teenager', was at some point 'a French structuralist Marxist', then simply a 'Marxist', and that politically 'I was with Lenin'.[56] Not only is Marx numbered among his influences but Hart also uses terms like 'class warfare', 'Marx's class contradictions', and 'Karl Marx nailed Victorian capitalism', observing that 'we are all deformed by class divisions'.[57] Elsewhere he states unambiguously that Marx's *Capital* 'written during the 1860s ... must still be the intellectual starting point for our own moment of world history, its imminent demise in the aftermath of the Cold War'.[58]

Alongside these endorsements, however, are found indictments of all things Marxist.[59] Thus, for example, Hart maintains that '[t]he Bolshevik revolution's violent repression of markets was a disaster', and argues for the necessity of transcending the 'binaries' of class.[60] Marx and Engels are chided for mistakenly thinking that workers are interested in the downfall of capitalism, an error he attributes to proletarians having non-class identities as a result of belonging to institutions outside the workplace. In support of this critique, Hart points out that such externalities are present as much in the informal economy as historically among Lancashire factory workers.[61] In his opinion, therefore,

---

55   An indication of this aporia is the positive reception given by Hart to the interpretation of labour regime change advanced both by David Graeber and by Jan Breman. For these endorsements, see Hart (2022: 30, 144) and Hart (2000: 439–443); on why they are misplaced, see Brass (2017b: Chapters 2, 7 and 16).
56   See Hart (2022: 2, 108, 142).
57   See Hart (2022: 2, 29, 57, 129–30), whose condemnation (Hart, 2022: 11) of the accumulation process follows that of Marx: 'Capitalism has always rested on an unequal contract between owners of money and those who make and buy their products. This contract depends on an effective threat of punishment if workers withhold their labour or buyers fail to pay up'.
58   See Hart (1992: 226), whose book about West Africa (Hart, 1982: 3, 15, 17,18, 40, 158, 170 n13, 182, 183) contains similarly positive views about the theoretical approach both of Marx and of Lenin.
59   This political ambivalence is reflected in his (Hart, 2022: 27) self-description as 'a radical conservative' plus the admission (Hart, 2022: 109) that when in America he 'applied to join the CIA'.
60   Hart (2022: 45, 74). Elsewhere 'Lenin's imperialism thesis' is described by him (Hart, 1990a) as 'passé'.
61   'Marx and Engels thought that the industrial working class would overthrow capitalism', declares Hart (2022: 137, 138), whereas 'Studies of the workplace must take in the

'Marx and Engels missed it all': quite simply, they were wrong. What they failed to understand was that working class collectivism is not incompatible with petty bourgeois individualism. In contrast with his supportive references to the presence and efficacy of class, Hart dismisses the latter concept: 'Marx and Engels' class analysis falls down immediately', since differences between proletariat, petty bourgeoisie, and lumpenproletariat are in his view non-existent. Distancing himself from class also has implication for the way alienation is conceptualised.

Endorsing Marx's concept of alienation and commodity fetishism, Hart notes that Marx 'captured the essence of Victorian capitalism in *Capital*', and that, as a result of having to 'work under conditions imposed by owners', people 'are estranged from their own humanity'.[62] Of central importance to Marx, however, was that alienation was class-specific: it was principally members of the proletariat who in economic terms could be said to be estranged, separated thereby from the value produced by their labour-power, a process which in turn generated the surplus-value appropriated by the capitalist. Before, during, and after this process workers obviously are – and remain – human: nevertheless, what is taken from them as a consequence of being alienated in this particular sense (the linchpin of Marxist theory about the way in which accumulation takes place) is not so much their humanity as the value that expenditure of effort and application of labour-power has produced. The multiple effects of this kind of aporia – for and against Marxism, for and against class – are difficult to avoid.

Self-Sufficiency

That post-Marxism licenses both from-below empowerment conferred by non-class identity, and its political legitimacy (= an authentic grassroots voice) is in keeping with the way the informal economy, the market, and humanism are all connected in the work of Hart. Although he accepts that the informal economy was hard for those 'unfortunate individuals forced to live in this way', the picture Hart draws of its denizens can be interpreted as a form of

---

62   institutions people devise for themselves outside it. The informal economy was a strategy of Lancashire factory workers too ... Marx and Engels missed it all. They clung to a contrast between working-class collectivism and petty bourgeois individualism that was never there ... Marx and Engels' class analysis falls down immediately. The lines demarcating the proletariat, petty bourgeoisie and lumpenproletariat dissolve'.
    Hart (2022: 48–49).

empowerment.[63] Challenging the negative image held by modernisation theory of the informal sector as a location inhabited by the destitute and unemployed, Hart (and others) drew attention to its positive side, composed of its economic dynamism. It was a place in which, through moonlighting among other things, it was possible to become a 'small scale entrepreneur', and where petty capitalism thrived.[64]

This is the reason why the idea of the informal economy as safety-net was received so favourably, both by governments and by international agencies and organizations concerned with Third World development, since it was now possible for them to characterize informal economic activity as a 'solution' to poverty, thereby avoiding the kind of reforms which not only entailed wealth redistribution and an attack on private property, but also the possibility of a socialist transition. In this respect, the official popularity of the informal sector as a panacea for urban poverty served the same ends as the Chayanovian peasant family farm, seen by many as an economically viable non-socialist answer to rural underdevelopment.[65] The seeming capacity of those in the informal sector to eke out a livelihood, and on occasion even to accumulate capital, conferred plausibility on arguments opposed to change, along the lines that its inhabitants 'are doing alright as they are, so there is really no need to change the economic system as presently constituted'. As easy to understand is why the concept appealed to those on the political right, who saw the informal sector activity as confirmation of their own economic (anti-taxation, off-the-books transactions, bourgeois individualism) and political (anti-state, anti-bureaucracy) values.

To his credit, and unlike other anthropologists or social scientists who initially get things wrong, Hart subsequently recognized that depicting informal sector dynamism as an alternative to modernisation ended up playing

---

63   See Hart (1973: 83, 84), who argues that 'for many urban wage-earners poverty is ever present, and [consequently] the informal sector provides opportunities of improving real incomes for this category as well as for the "jobless"'.

64   The challenge to modernisation theory was expressed thus by Hart (1973: 61): 'Does the "reserve army of urban unemployed and underemployed" really constitute a passive, exploited majority in cities like Accra, or do their informal economic activities possess some autonomous capacity for generating growth in the incomes of the urban (and rural) poor?'. This challenge is repeated and reinforced subsequently (Hart, 1973: 81, original emphasis): 'It is generally understood that growing residual underemployment and unemployment in the cities of developing countries is "a bad thing". But why should this be so? In what way precisely does this phenomenon constitute a *problem?*'.

65   See Chayanov (1966; 1991).

into the hands of those opposed to reform.⁶⁶ In this book, therefore, his later comments reflect this outcome, taking a more critical view of the informal economy, a contrast with earlier perceptions of it as 'autonomy' and in some sense empowering.⁶⁷ Ironically, this revives the aspect of the informal sector which Hart originally questioned: the presence there of the industrial reserve on which capitalism draws, and on which the capacity of the accumulation process to reproduce itself ultimately depends. It also reactivates the negative view of the informal sector held by Marxism, which sees it as disempowering in terms of class struggle and a transition to socialism. This in turn points to a continuing difficulty in the work of Hart, one that can be seen as a twofold legacy: his espousal of humanism, and his earlier positive characterization of the informal sector, each of which inform a pervasive adherence to the view that the market is efficacious.

Humanity's Priority

An additional instance in the work of Hart of the world reflecting the experiences of the self is his support for the idea of movement-as-liberty, which combines humanism, liberalism, and the market.⁶⁸ Just as he himself has travelled

---

66 'I had no excuse', confesses Hart (1992: 220), 'I had lived in a slum for two years; fieldwork had given me a chance to see what was going on, even to take the side of the people. It would be nice to say that the anthropological method is inherently democratic and superior to the remote speculations of academic bureaucrats; and so it ought to be … But I was so anxious to get the big picture, to go with the power and join the bureaucracy, that I transformed my fieldwork into a gimmicky idea that development economists were able to absorb into their Panglossian vision of the world'. This sort of *mea culpa* is not merely welcome for its honesty, but also and sadly a rare – indeed, very rare – occurrence, one not usually encountered in social scientific discourse.

67 See Hart (2022: 191ff.). Hence the following: 'It seemed now that the informal economy had gone global through a combination of neoliberalism and the digital revolution … the idea of an informal economy was born when the post-war era of development states was drawing to a close. The 1970s were a watershed between … state management and the free market … The informal economy started off as a way of talking about the Third World urban poor [but its] improbable rise to global dominance is an outcome of the mania for deregulation since 1980'.

68 Studying the street economy in Ghana, Hart (2022: 2) makes clear its compatibility with his own experience: 'I found that I had considerable knowledge of markets and money. The "informal economy" was a free market zone, operating outside the law. I felt at home in it'. Earlier much the same view is hinted at elsewhere (Hart, 1973: 88–89): 'Socialists may argue that foreign capitalist dominance of these economies determines the scope for informal (and formal) development, and condemns the majority of the urban population to deprivation and exploitation. More optimistic liberals may see in informal

the globe in search of work, so is his contention that everyone, everywhere, should be permitted to do the same: cross-border freedom of movement is a process that in his view ought to be both universal and international.[69] Because '[t]he world belongs to us all', observes Hart, 'we should be able to move as we wish', adding: 'Global capital will only be checked when most people are free to move'.[70] Arguing that 'freedom of movement ... should be reinstated as humanity's priority in the drive to make a viable work society', Hart makes clear the close link between his humanistic vision and its realization in the form of freedom of movement across the globe as a 'human right'.[71] Equally clear is where the obstacle to this realization lies, since he contends that 'territorial states maintain inequality by restricting the movement of people from poor to rich areas'.[72]

Given the positive view Hart takes of the freedom of movement, as unsurprising is his view that 'the Brexit referendum was fuelled by anti-immigration feeling and resentment of economic decline, [and] was followed by significant right-wing gains in the 2019 [UK] election [when] blaming foreigners included leaving Europe', the result being that places which previously endorsed social citizenship were 'now consumed with neoliberalism's identity politics'.[73] Hart asks, rightly, 'where does the "immigration problem" come from?', and answers, also correctly, 'impoverishment [since] Europe's "refugee crisis" is caused by African poverty' among other things.[74] Agreeing that 'Western workers face increased competition at home and abroad', he downplays its impact, noting simply that this has been so for the past century.

---

activities...the possibility of a dramatic "bootstrap" operation, lifting the underdeveloped economies through their own indigenous enterprise'.

69 This connection between self and the world is underlined by Hart (2022: 38–39, 116) in the following manner: 'Kant [one of his influences] held that ... the basic right of all world citizens ... should rest on conditions of universal hospitality. A stranger should not be treated with hostility when they arrive on someone else's territory. We should be free to go wherever we like in the world, since it belongs to us all equally'; and 'I had found liberty in America. I had the chance to spread my wings. The idea of freedom runs deep in the United States and is synonymous with movement'.

70 Hart (2022: 172).

71 Hart (2022: 177).

72 Hart (2022: 172). 'A new free trade campaign would dismantle the institutions of privilege by insisting on on movement as a human right', claims Hart (2022: 177): 'Its motivation would be liberal – an attack on global divisions in the name of our common humanity ... [r]emoving state jurisdiction over international movement is ... essential now'.

73 Hart (2022: 171).

74 Hart (2022: 174–75).

It is true that the 'immigration problem' is a result of poverty in the sending nation, but – ironically – it also derives from its opposite. Inhabitants of poor countries who possess skills (due in some instances to foreign aid programmes) see little or no economic future where they live for people like themselves, so in a neoliberal global context they opt to sell their labour-power elsewhere. Migrating to where pay is higher and work conditions better, such free movement deprives the sending country of badly needed skills, consolidating poverty there, contributing thereby to what has been called the 'development paradox'.[75] In the receiving nation the same free movement contributes to an expanding industrial reserve, generating thereby the acute labour market competition that fuels a populist backlash, undermining any existing consciousness, political solidarity, and struggle based on class.[76]

Restlessness

Again ironically, the very process of which Hart disapproves – 'anti-immigrant feeling' leading to 'right-wing [electoral] gains' – has been fuelled by the kinds of processes he regards as positive. Namely, the centrality of the market and the desirability of free movement which, when it applies to migration in the form of open-door policy, generates acute labour competition that advantages employers and disadvantages those in the receiving country who aspire to or already have jobs. This in turn generates the identity politics and 'anti-immigration feeling' of which he (and this writer) correctly disapproves. Deregulating the labour market, by removing controls on cross-border freedom of movement, plays directly into the hands of capital, and rather than eliminating inequality – as Hart maintains – would much rather perpetuate this, by

---

75 A 2021 report by the UK House of Commons indicated that there are more health professionals of Ghanaian origin working for the British NHS than there are in Ghana itself. This despite the fact that Ghana is on the World Health Organization red list, which forbids the active recruitment of its health workers by other nations.

76 As has been argued elsewhere (Brass, 2021b: Chapter 8) it is possible to interpret the debate about Brexit as the clash between rival populisms, one for and one against immigration. What postmodernism has done is to move into an ideological space occupied historically by populism: to the postmodern argument emphasising the cultural identity of the migrant-as-'other'-nationality, far right populism counterposes an argument similarly emphasising cultural identity, only this time the nationality of the non-migrant worker (i.e. British selfhood). In the absence of socialist ideas/practice, and as capitalism spreads across the globe, this form of nationalist discourse can be deployed effectively by populists who claim it is the only way to safeguard/retain workers' jobs and living standards.

undermining both local and migrant alike, leading over time to a decline in the living standards of each.

Notwithstanding his opposition to neoliberalism, contending rightly that 'counter-revolution got the world moving – in the wrong political direction', Hart seems not to recognize the extent to which his support for movement-as-liberty is compatible with (and even central to) neoliberal theory about the labour market.[77] In part, an inability to spot the presence of this overlap stems from his ambivalence about Marxism; its consequence in theoretical and political terms is a depriviledging of the link between on the one hand freedom of movement on a global scale and a burgeoning industrial reserve army of labour, and on the other the impact of migration into metropolitan capitalist nations, in the form of enhancing the acuteness of labour market competition leading to a populist backlash. It also makes difficult an understanding not just how movement-as-liberty or free-movement-as-a-human-right negate the ability of any future socialist government to plan and regulate both the labour market and the economy, but also how the labour market is itself transformed by capitalist development on a global scale, becoming as a result a Hobbesian war of all against all.[78]

Central to 'human economy', therefore, is 'a world of movement' – a perpetual labour market competition across borders, the very process of which neoliberalism approves and on which its capacity to accumulate depends.[79] In the end, the support expressed by Hart for individual autonomy is not so different from the backing neoliberalism gives to the same ideology. Hard to avoid is the view that, espousing as he does a quasi-libertarian approach to individualism, movement, money and the market, Hart appears in a fundamental sense to have bought into a version of the neoliberal project. This

---

77 Hart (2022: 128).
78 Notions of the market as essentially benign are dispelled by the terrifying images of social breakdown portrayed by Adam Curtis in his 2022 documentary film *Russia 1985–1999: TraumaZone (What It Felt Like to Live Through the Collapse of Communism and Democracy)*. It charts the impact of 'shock therapy' inflicted on the Russian people as a result of the imposition of the market economy. A combination of privatisation, deregulation, the lifting of price controls and strict austerity led not just to increased corruption and inequality, manifested in high levels of illness and mortality, but also to a catastrophic general decline in living standards: whereas only two per cent of Russians were classified as poor in 1987–88, by 1993–95 this figure had risen to 50 per cent.
79 At times the concept 'development' as used by Hart (1990b: 3) – 'It is a characteristic of our epoch that people should seek a better life ("development") and that their condition should be one of restlessness' – seems indistinguishable from the way the same term is understood by neoliberals.

in turn can be traced back to his fieldwork in Accra: because the informal economy 'worked', so to speak, its smooth functioning as a 'free market zone' prompted an interest in the role and importance of money and markets. As Hart himself puts it, coming across the informal economy, he concludes significantly, 'fed my Manchester liberal side and temporarily modified the socialism of my youth'.[80]

Conclusion

Revealing and informative, the de facto autobiography by Keith Hart covers an eight-decade period when erstwhile secure academic employment declined while debate in the social sciences about the desirability (or otherwise) of Third World development shifted dramatically. Not the least interesting aspect of his narrative is that the career trajectory it depicts – an agglomeration of numerous part-time/temporary jobs, dispersed worldwide – prefigures what has now become an increasingly common pattern in the university system of western capitalist societies. Inescapable, too, is the extent of the contradictions informing the self that is Hart: an influential and knowledgeable scholar who has largely turned his back on academic institutions; an anthropologist who doesn't much like anthropology as a discipline; an accomplished fieldworker who disapproves of fieldwork; an opponent of neoliberalism and identity politics who supports free movement that fuels job competition and populism; and a politically eclectic leftist who regards money and the market as positive.

Disagreement with aspects of his intellectual approach – ambivalence about Marxism, espousal of the market – should not detract from what Hart has managed to achieve, frequently in defiance of adversity (health, peripatetic employment). He has not only contributed substantially to the formation of a major concept – the informal economy – in the study of development, but also has had the courage to admit subsequently why his original interpretation might have been problematic. As significant has been his contribution to the task of popularizing (in the best sense) anthropology, taking it away from the

---

80 Hart (2022: 86). The clue to the importance of this shift in political emphasis from Marxism to liberalism is made explicit right at the start of the narrative: 'The tension between belonging to a collective and individualism is a central theme of [the] book', accepts Hart (2022: 2), who then declares that '[a]fter a long period as a Marxist, I immersed myself in classical liberalism'.

confines of the university department and out into the wider world (where it truly belongs). More deserving than most of a *festschrift* volume, Hart has thus far not received one. In a sense, however, he doesn't need this kind of peer-group recognition, since his record as set out in this important book speaks for itself.

CHAPTER 8

Labour Regime and Development: Deproletarianisation and Neo-bondage Compared

> Many latter-day accounts on developmentalism lack historical depth.
>> A critical observation by JAN BREMAN that, in the light of his own problems constructing a history of the debate about development, is indeed curious.[1]

∴

Introduction: Explaining Unfree Labour

In one way or another, unfree production relations – their characteristics, their similarity-to/difference-from free equivalents, their coercive/non-coercive character, their historical/current presence/absence, their compatibility/incompatibility with capitalism – have been a central aspect of post-war debate in the social sciences generally, and especially in the field of development studies. The political and economic role of labour-power that is unfree has been an important feature of academic disputes about a wide range of current issues, extending from Brexit, the rise and consolidation of populism and the far right, to the expansion in the world industrial reserve army, the structure of the global labour regime, the formation of and struggle over worker consciousness, and the difference between and desirability of class as distinct from non-class identity.[2]

Accordingly, there can be no doubt regarding the importance of this debate for development studies. Where unfree production relations have been reintroduced into a labour process that is capitalist, political economy teaches that such an event corresponds historically to a regression. This is because so much – politically, ideologically and economically – hangs on whether or not

---

1 Breman (2024: 351).
2 On the connection between these socio-economic phenomena, see Brass (2017b).

worker emancipation has been achieved. In the context of class formation/ struggle, workers who are free possess economic, ideological and political advantages that unfree counterparts do not have. These include the capacity to bargain over issues such as the sale and price of their labour-power, and the conditions and duration of work: indeed, how else is one to explain the acceptability to employers of unfree labour. The presence of labour-power that is unfree also feeds into discussion about the link between globalization and a burgeoning industrial reserve army, where it features as popular concern regarding the economic impact that migration from Third World countries has on labour market competition in metropolitan capitalist nations, and the rise there of hostility that fuels populism.

Examined here, therefore, are the arguments made by what are currently two of the main interpretations (deproletarianisation, neo-bondage) about unfree production relations, the way these are defined and why, together with the connection – if any – to the accumulation process. Of the reasons why this debate is of interest, three in particular should be mentioned. First, the most recent contribution to this discussion is by Jan Breman, who formulated the term neo-bondage, and whose latest book constitutes an attempt to locate this concept in its historical, epistemological, and political context. Second, because not only are the two approaches the same or different, but also why this is so. The latter aspect is politically significant, in that deproletarianisation perceives a socialist transition as a solution to the problem of unfree labour, whereas neo-bondage posits a remedy achievable within capitalism. Whereas one interpretation is linked to a progressive systemic outcome of the development process, the other is not.

And third, recent intervention in this debate by geographers indicated that one of these interpretations – the deproletarianisation approach – might be useful in addressing questions about changes in the capitalist labour regime that are of importance to the discipline.[3] Because of this, provided below is the background to the debate, the focus of which is on both the definition and the acceptability (or otherwise) to capitalism of production relations that are unfree, since it is these aspects which are most at issue in the two interpretations under consideration. Consequently, an object here is to bring this aspect of the debate on the capitalism/unfreedom to the attention of

---

3 The intervention was by McCusker, O'Keefe, O'Keefe, and O'Brian (2013), who outlined the case that geographers ought to engage with the deproletarianisation approach. The accuracy of the analysis on which their argument is based has been acknowledged elsewhere (Brass, 2022c, 2023e).

geographers – especially those engaged in the study of Third World development – who have shown an interest in contributing to this discussion.

The presentation which follows consists of two sections, the first of which examines the two paradigms, in terms of their epistemology and consistency, whilst the second considers in more detail how and why free/unfree production relations are linked to the process of capitalist development, together with theoretical problems arising from this, and the political implications of each paradigm.

I

Deproletarianisation, Neo-bondage

Deproletarianisation refers to the acceptability to capitalism of labour-power that is not free, whereby agribusiness enterprises, commercial farmers and rich peasants reproduce, introduce or reintroduce unfree relations. Such workforce restructuring involves replacing free labour with unfree equivalents, a procedure frequently resorted to by employers. This labour process decomposition/recomposition is itself a concept based on Marxist theory about class struggle. Accordingly, the object of deproletarianisation is to discipline and cheapen labour-power, an undeniable economic advantage in a global context where capitalist producers have had to become increasingly cost-conscious in order to remain competitive. Following on from the adoption of the Green Revolution in the Third World, coupled with labour market deregulation in metropolitan capitalist nations, workforce restructuring benefitted from the capacity of producers to draw on a burgeoning industrial reserve army.

Central to this class struggle approach is that, historically and currently, deproletarianisation is characterized by multiple shifts: in terms of its systemic operation, in unfreedom itself, and in the way it is enforced. Rather than being confined epistemologically to pre-capitalist modes of production, and as such destined to be replaced by labour-power that is free, deproletarianisation underlines the capitalist nature of production relations that are unfree. With regard to the latter, therefore, unfreedom as a result of indebtedness has moved from being an aspect simply of permanent employment to a condition also of seasonal and casual equivalents; from just locals it extended also to migrant labour; and from male to include female workers. Beck-and-call arrangements covering permanent/seasonal/casual workers who are unfree enables them to work for other employers only when their creditor-employer did not himself want their labour-power. An effect of withholding wage payment due meant also that a bonded migrant labourer was and is compelled to return and continue working for the same employer the following season. In terms of enforcement, unfreedom involved coercion exercised horizontally by the kin/caste members of the indebted worker, not just vertically by the landlord or merchant.

Where defining characteristics and purpose are concerned, it is virtually impossible to distinguish deproletarianisation from Breman's conceptualisation of neo-bondage.[4] The latter is said not only to be compatible with capitalism, but also differs from pre-capitalist (= 'traditional') forms of unfreedom in length of employment, applying as much to those seasonal and casual workers, male and female, bonded by debt.[5] Similarly, neo-bondage itself stems from procedures such as advanced payment, borrowing, or withholding payment, and in a capitalist agriculture its debt-servicing work obligations are enforced from within the kinship domain of the bonded labourer.[6] That neo-bondage has been influenced by the deproletarianisation approach, to the extent of adopting many of its arguments under a different label, is an issue difficult to avoid. Again, just two examples will confirm this. First, the concept 'dispossession' is extended by Breman to cover the deprivation of the capacity by a worker to sell his/her own labour-power, noting that Harvey could have done this but failed to do so, reproducing thereby an earlier and similar critique.[7] And second, Breman describes as 'my thesis' the argument 'that the loss of control over their own labour power [is] a final step in the trajectory of dispossession

---

4 In my review of a book *Chains of Servitude*, edited by Patnaik and Dingwaney, published in the October 1986 issue of *The Journal of Peasant Studies*, 14/1, pp. 120–26, is found the following on page 124: 'This change entails not just the monetization of bondage but (perhaps of greater importance) a precise relocation of the relationship in terms of the labour process. The transformation corresponds, in short, to a shift in the immobilizing function of debt from a continuous and intergenerational basis to a more period- and context-specific basis; its operationalisation is as a result confined to the months of peak demand in the agricultural cycle, when sellers of labour-power would otherwise command high prices for their commodity'. Breman would have been familiar with this, since in his contribution to a special issue of that journal (Breman & E. Valentine Daniel, 'Conclusion: The Making of a Coolie', *The Journal of Peasant Studies*, 19/3–4, 1992) he cites that very same book review.

5 See Breman (2023: 159): 'The new bondage differs from the traditional one in terms of the short duration of the contract – often no longer than for one season; its less inclusive character – just labour instead of a more encompassing beck-and-call relationship; and finally, its enhanced vulnerability to non-compliance. Again, it is the labour power which is required by working off the amount of debt'. On the shift of unfreedom to seasonal workers, see also Breman (2023: 73, 82).

6 Breman (2023: 35, 36, 150). That the enforcement of debt bondage is now carried out by the labour contractor 'who belongs to the same milieu as the worker', and, where family labour is concerned, by the household head are arguments first encountered in Breman (1999b: 469).

7 Compare Breman (2016: 255–256) and Brass (2011: 148).

among the landless workforce', a characteristic of neo-bondage that reproduces more-or-less exactly the same description of deproletarianisation.[8]

Of additional interest, therefore, is the sequence in which the two interpretations emerged. The case linking unfree production relations to capitalism, together with the characteristics associated with deproletarianisation, was already in place during the 1980s, appearing in a variety of publications.[9] The term neo-bondage did not surface until 1993, when it was mentioned in the new edition of the 1974 book about Gujarat.[10] Thereafter it sprang into life, as a concept explaining the presence of a new form of unfree labour that was compatible with the model of capitalist development occurring in that part of India, since when it has featured in everything Breman has published on this issue.[11] This is not so surprising, given that until the advent of neo-bondage it was still being argued that debt-induced unfreedom was not – and could not be – an aspect of capitalist agriculture.

On occasion, moreover, closely following the lead of deproletarianisation has required a change of mind.[12] Because its focus is on class struggle as the cause of unfree production relations, the deproletarianisation approach also questioned the premiss that saw unfreedom as an effect simply of labour shortages, an argument made by Nieboer with regard to slavery.[13] Having initially endorsed the view linking unfreedom to labour shortages ('I still endorse

---

8 For this categorization of neo-bondage see Breman (2016: 199), which can be compared to the following description of deproletarianisation (Brass, 2011: 273): 'It is precisely this ownership, exercised by workers over their labour power, that an employer has to deprive them of so as to exert in turn full control over a modern production process. Deproletarianization captures this fact, by underlining the extent to which it is necessary for capital to close off even this limited economic autonomy. It recognizes conceptually that what has happened to workers is that, finally, they have become what Marx said they would: men and women of no property. Capital has in short taken from them their sole remaining property'.

9 These articles include Brass (1980, 1983a, 1983b, 1986a, 1986b, 1988, 1990) and Brass & Bernstein (1992).

10 The earliest discernable reference to the term 'neo-bondage' is in Breman (1993: 300) where it is stated: 'On the basis of both historical and contemporary research, I ascribe to the thesis that the neo-bonding of labour need not conflict with the capitalist mode of production'.

11 See in particular Breman (1996: 167–69).

12 Thus, for example, in the mid-1980s, Breman (1985: 311) states: 'I shall regard as unfree only that form of debt-labour which is rooted in non-economic coercion.' Four decades on, however, he argues the opposite, insisting (Breman, 2023: 159) that 'the unfree relationship does not have its origin in extra-economic coercion'.

13 On slavery as the outcome of labour shortages, see Nieboer (1910). Linking unfreedom to a shortage of workers, as does Nieboer, was a theory questioned by me (Brass & Bernstein, 1992: 17).

Nieboer's original premise that claims a shortage of labour produced by a high land-to-man ratio is a crucial ingredient leading people to force others to work for them'), Breman then dismisses the attribution of the neo-bondage unfree production relation to labour shortages, and replaces it with the one advocated by theory about deproletarianisation.[14] He not only persists in ignoring this *volte face*, but – perhaps as a result – continues to argue even now that his views about the labour regime developed against the grain, rather than as is the case in step with existing orthodoxy.[15]

Unfreedom, Patronage, Politics

It is impossible to understand either the significance of the capitalism/ unfreedom debate or the place in this of rival concepts and interpretations without reference to the prevailing orthodoxy in the 1960s development decade: namely, that as capitalism spread, unfree production relations would be replaced by free equivalents. This view, associated with paradigms such as modernisation theory and the semi-feudal thesis, informed the early work of Breman, where it took the form of positive endorsements of concepts such as 'patronage' and 'subsistence guarantee' applied by him to bonded labour – the *hali* system – in the Indian state of Gujarat. In what is the sole reference in the book to the interpretation of unfree labour as deproletarianisation, Breman complains that it unjustifiably accused him of portraying debt bondage as having a positive side, dismissing such a view as 'a grotesque misrepresentation of [his] writings on the subject'.[16] Evidence suggests otherwise: far from being

---

14 For the endorsement of the labour shortages, see Breman (1993: 298). For the opposite view taken up subsequently, see Breman (2023: 152, 154), where he states: 'Having rejected standard explanations for labour migration which assumed lack of adequate manpower ... my proposition was that workers imported from elsewhere were both cheaper and allowed for greater command over their labour power'.

15 Hence the following view (Breman, 2024: 35–36, emphasis added): 'The proletarianization of the low-ranking peasantry that accompanied this reform was achieved under Dutch authority and policy. This led me to focus my attention on the landless class that is largely absent in most publications on what was and remained an agricultural economy. An increasing mass of dispossessed people was forced to depend on hiring out their labour power. Often condemned to a life in bondage, this class– embodied in the figure of the colonial coolie – would become a recurring subject in my further empirical research. Discussions on this theme, usually in historical journals, gave rise to my involvement in polemical debates on the interpretation of colonial policy and the extent to which *the views I expressed contradicted the established wisdom*'.

16 For this complaint, see Breman (2023: 101). The section to which he objects (Brass, 1990: 41), dealing with his characterization of bonded labour as a form of subsistence

'a grotesque misrepresentation', this critique is much rather an accurate representation of his views.[17]

That patronage and its debt bondage relation was – and to some degree is still – seen by Breman as having a benign dimension is clear, not least from the observation that 'the seigneurial code of conduct' as embodied in this arrangement 'stood in marked contrast to the harsh treatment ... meted out to landless labour in the late colonial era'. Patronage was not all bad, it seems.[18] In keeping with this interpretation, at that mid-1970s conjuncture he accepted that the *hali* system lacked freedom, but nevertheless downplayed its coercive element and instead emphasised strongly the benign and mutually beneficial aspects of the relation. 'It is therefore doubtful', he maintained, 'that the *hali* strove to end attachment', concluding unequivocally that 'servitude was sought rather than avoided by the [debt bonded farm servants]'.[19] The same kind of benign interpretation structured his view about the nature of the subsistence guarantee.

---

guarantee, merely repeats a complaint made earlier (Breman, 1993: 297, note 1). For the same critique made by him subsequently, see Breman (1996: 168–69).

17   That in the early 1960s bonded labour was perceived by Breman as benign is confirmed by the fact that its subjects were thought by him not to struggle against this relational form. Of a colleague with whom he discussed worker unfreedom, therefore, Breman (2024: 391, emphasis added) observes: 'Wim heard that his host employed farm labourers as bonded servants. He not only criticized this man and other large landlords but also made it abundantly clear that such practices were morally and socially reprehensible. These members of the elite listened to Wim in silence, but his opinions remained the main topic of discussion for many days after he had left. *He had also responded with incredulity when, in answer to his repeated questioning, I told him that I had found no evidence that the landless proletariat resisted bondage.* It was impossible, he insisted, as slaves have always and everywhere risen up against their masters. His critical comments kept me preoccupied. *In the historical source documentation that I consulted extensively after completing my fieldwork, I also found no evidence of any organized proletarian resistance*'. In an accompanying footnote, he confesses that 'this conclusion later proved incorrect, when I discovered an archive with factual evidence of a protest movement in the late-colonial period. I incorporated this revised view into a new [2007] publication'.

18   Breman (2023: 122). 'By patronage', he writes (Breman, 1974: 18ff.) at that conjuncture, 'I mean a pattern of relationships in which members of hierarchically arranged groups possess mutually recognized, not explicitly stipulated rights and obligations and preferential treatment'. He continues: 'A member of a low caste who was a follower of a prominent patron was assured of security and safety in times of scarcity and tension'. Accepting that 'continuous bondage ... operated against the farm servant', Breman (1974: 21) maintains that 'even in such cases, complete exploitation and excessively arbitrary treatment were contrary to the interests of the landlord, who had to be sure of the support of his followers'.

19   See Breman (1974: 43–44, 45), who elsewhere writes (Breman, 1989: 151, 152) in a similar vein both that '[a]s is evidenced by the extent of the benefits enjoyed by the hali, he received total care ... he depended for his every need on what the master put into his

The subsistence guarantee argument surfaces regularly in the early volumes by Breman, and from the many examples just two will suffice to underline his adherence to this concept. Noting that 'it is assumed that bondage was simply imposed', and that '[t]his is how the hali system has usually been represented [whereby great] emphasis has been laid on the coercion applied by the masters, both in contracting and in maintaining the relationship [but as] an interpretation which takes nothing else into account [it] is too one-sided', Breman concludes unambiguously that 'it is difficult to maintain that the servitude of the ... agricultural labourers was forced upon them against their will. On the contrary, for lack of continuous employment ... those who had managed to find someone to provide for them had every reason to consider themselves lucky'.[20] It is clear, therefore, that initially Breman regarded patronage and subsistence guarantee in positive terms, maintaining that attached labourers preferred servitude, and considered themselves 'lucky' to be employed thus, that bondage was for them an economically attractive proposition in which the element of debt was unimportant. Such an interpretation throws into question his current view that he perceived unfree relations in wholly negative terms.

Because of this epistemological link between semi-feudalism, patronage, and debt bondage across all his early publications – spanning not just 1974 and 1978 but also 1985 and 1993 – Breman was still convinced until after the latter date that '[t]o continue to define the present relationship between farmer and worker in terms of bonded labour creates problems in my view', and further that '[b]ased on my own research ... it is my opinion that these features [low remuneration, indebtedness] have more to do with exploitation than with bondage'.[21] In turn, this link feeds into the sort of political solution envisioned by him as a way of resolving the problem of unfreedom.

In contrast to the deproletarianisation approach, which connects the eradication of unfree labour to a socialist transition, the solution to neo-bondage envisaged by Breman is located within capitalism itself: namely, that the

---

hands', and that consequently 'the element of patronage prevented any pronounced tyranny and complete exploitation'.

20 See Breman (1974: 43, emphasis added). Much the same positive view about the 'subsistence guarantee' can be found in a later book, where he (Breman, 2007: 37, 48) states: 'The economic security the relationship offered [the indebted worker] explains why the [farm servants] preferred to bond themselves to a master in this way. Reports that the *halis* were better off than [those] who had to make a living as unattached workers must be viewed in this light [which forces] us to modify the interpretation of halipratha as a form of labour bondage', and further, 'the need for food security was the primary motive for the [worker] to enter into a relationship based on bondage and to see their master as a benefactor'.

21 Breman (1993: 309, 310–11).

neoliberal capitalist state *ought* to care more than it does about the plight of workers, by passing and then enforcing legislation.[22] This is the kind of solution that entails a return to a 'caring'/'kinder' capitalism and its state, one that leaves capitalism as a system intact. Herein lies a major teleological and political difference between the neo-bondage framework (with its the capitalist-state-ought-to-care solution) and that based on deproletarianisation (for which the solution is a transition to socialism).

What this adherence to the capitalist-state-ought-to-care approach reveals is a misunderstanding of the systemic forces at work. It is not, as Breman supposes, a question of the capitalist state choosing to do, or not to do, something or other, but the inexorable economic logic of the accumulation process imposing itself. As Milton Friedman, the high priest of *laissez-faire* economic theory observed half a century ago, apart from making profits for its shareholders, the market does not recognise an obligation to anyone or anything, noting that 'discussions of the "social responsibilities of business" are notable for their analytical looseness and lack of rigor'. Yet Breman attributes to capital, and also to its state, an ethical dimension that neither in fact possess.[23]

An additionally problematic aspect of this capitalist-state-ought-to-care narrative concerns the wider systemic outcome resulting from replacing a Marxist agenda with a non-Marxist one. Hence the corollary of attaching a non-Marxist political agenda (linked to the neo-bondage approach) to what was originally a Marxist one (linked to theory about deproletarianisation) is that the longer a socialist transition is postponed, by not putting this leftist objective on a political agenda, together with that of migration – free and unfree – and the industrial reserve army, the more workers will in periods of crisis move towards reactionary populist solutions which seemingly offer to protect their jobs, culture, and livelihoods. In a sense, therefore, Breman has still not broken with the concept patronage, in that when considering the role of the capitalist state he returns to this earlier notion, albeit now with a different subject: instead of the landowner, therefore, it is the capitalist state which in his view ought to be the provider and source of patronage.

---

22   Hence the view (Breman, 2023: 113–14, see also 119) that 'while employers denied any responsibility to provide even an elementary quality of life for their workers, the state showed itself unwilling to ensure and enforce the social reproduction of the rural proletariat at a minimal level'.

23   This issue is considered in greater detail elsewhere (Brass, 2015b).

II

Differences Explained?

Anyone hoping to learn from his latest book why Breman changed his mind – so abruptly and completely – about the accumulation/unfreedom link, shifting from regarding debt bondage as incompatible with capitalist development to the opposite view that bonded labour is its preferred form of production relation, together with the reason for the emergence as if from nowhere of his concept neo-bondage, is in for something akin to a magical mystery tour.[24] This is because what the reader encounters is in effect the unfolding of a chronology, the undisguised agenda of which is to demonstrate that he himself did indeed recognize earlier than thought the case regarding the acceptability to capitalism of labour-power that is unfree. Accordingly, structured by a comparison of what fieldwork in Gujarat revealed over two conjunctures – the early 1960s and the late 1980s – a discernible and interrelated sub-text emerges: namely the claim by Breman that he has always recognized the acceptability to capitalism of unfree labour; that the concept neo-bondage accompanied such recognition; and that, consequently, any criticism both that he has changed his mind about this, and that neo-bondage emerged only much later, is incorrect.

Preparing the ground for this argument, that as the link between landlord and unfree worker 'was structured by capitalism', the acceptability to the latter system of production relations that are not free is something he himself has always recognized, and thus an issue about which he did not change his mind. Noting that '[m]y study zooms in on the origin of labour bondage in peasant economies which have dominated the subcontinent of south Asia from an unrecorded pre-colonial past until the post-colonial present', Breman states that the connection between 'landowner and landless became a relationship structured and cultured in a capitalist mode of production'.[25] This same claim is then repeated and reinforced at a number of subsequent points in the book.[26] By arguing that Thorner regarded unfree labour-power and capitalism as incompatible, and then stating that it is not a view that he himself holds, Breman seems once again to create a space for the position that, unlike Thorner, this very link is one he – Breman – recognized right from the outset.[27]

---

24 A noticeable contrast exists between on the one hand an earlier, and because more self-critical a welcome, acceptance by Breman (2013: XII) of his 'inconsistency in the argument put forward', and on the other the doubling down by him on problematic claims and arguments in the latest volume.
25 Breman (2023: 30).
26 Breman (2023: 122).
27 Breman (2023: 120).

Accepting that his original judgment – the outcome of a decline in attached labour was 'a shift from unfreedom to freedom' – may have been wrong, Breman observes that as a result 'I was soon compelled to abandon this idea'. As to what lay behind this change of mind, according to him it was because he himself quickly realized his own initial view was incorrect, and that consequently 'a more nuanced interpretation ... was required'.[28] Far from he himself recognizing this error and rapidly correcting it, however, Breman continued to insist that the change entailed the replacement of unfree labour relations with free equivalents. Indeed, when it was suggested that his findings might be explained better as a case of deproletarianisation, the latter was dismissed as an alternative interpretation defended 'with more obstinacy than plausibility'.[29]

In support of the view that he has recognized as neo-bondage the compatibility between capitalism and unfree labour all along, Breman invokes the findings of his fieldwork conducted on sugarcane production in Bardoli in the late 1970s and the mid-1980s.[30] 'I pointed out', he says, 'in various publications that "a capitalist mode of production ... by no means precludes certain forms of absence of duress, emanating for example from the necessity to enter into debt"', the reference being to a late-1970s two-part article where these findings were presented.[31] Placing the conceptualisation of the acceptability to capitalism of unfree labour as neo-bondage at the same late-1970s conjuncture is a view repeated throughout the narrative.[32] These claims are faced with a

---

28 Although '*hali pratha* had disintegrated in the preceding decades', notes Breman (2023: 72), 'features of the past had clearly lingered on ... In my initial understanding, I tried to explain the changed nature of their relationship in terms of the disappearance of ongoing attachment, a shift from unfreedom to freedom. This would have implied that although many Halpatis remained employed as farm servants, they were no longer tied down in bondage. However, I was soon compelled to abandon this idea. The suggestion of a straightforward, abrupt transition of redemption from attachment was definitely a misconceived appraisal of the changes that had taken place. A more nuanced interpretation of what had gone from and what remained in their relationship was required. The plight of indebtedness which prevailed among nearly all of them [Halpatis] had not withered away'.

29 See Brass (1997b) for deproletarianisation as an alternative explanation, a suggestion rejected by Breman (1999a).

30 By the early 1970s, labourers no longer considered themselves 'at the unrestricted disposal of their employers' (Breman, 2023: 32); that is, they were not subordinated, and thus in keeping with the characteristics of neo-bondage.

31 See Breman (2023: 162). The two-part article, in succeeding issues but published in different years, is Breman (1978/1979).

32 See Breman (2023: 125, 144, 155–56). 'Before completing the [1985] monograph which attested to the main gist of my findings', argues Breman (2023: 123, emphasis added), 'I wrote up this major operation which big business instigated and managed from beginning

number of difficulties, ones that call into question the related arguments that, as he already recognized the capitalist nature of unfree production relations at that conjuncture, the concept neo-bondage was itself already in place then.

To begin with, that same quotation continues thus: 'It also explains why there are certain features in the plight of the cane-cutters which ... *are essentially those of pre-capitalist relationships*'.[33] Rather than being evidence for his connecting unfree labour and capitalism, therefore, the full quotation reveals that at that point he still regarded unfreedom as a 'pre-capitalist labour relationship'. Moreover, a crucial difference exists in the way the Bardoli labour regime itself has been interpreted. In his initial 1978/1979 two-part article, therefore, Breman characterized labour contractors as benign, debt bondage as a feudal/pre-capitalist relation, migrants as being 'free in their choice of where to go', and capitalism as necessitating 'a class of free and unattached wage-earners'. In later books, however, his account of the same capitalist labour regime at that same conjuncture has altered substantially, and is now presented as an example not of 'free and unattached wage-earners' but rather of 'neo-bondage'.[34]

More importantly, it is clear that even in the mid-1980s – seven years after the publication of that two-part article containing the Bardoli case study – Breman was still arguing that unfree labour was on the decline, and being replaced by free equivalents.[35] Indeed, on this particular issue he was

---

to end in a separate and earlier published essay (Breman 1978). *It turned out to be the stepping stone for my conceptualization of neo-bondage*'.

33 See Breman (1978-79: 190, emphasis added).

34 For later interpretations of that same Bardoli fieldwork, see Breman (2013: 336ff.; 2019: 196). Despite containing mention neither of neo-bondage nor of unfree labour, an article by Breman (1990) dealing with the role of the labour contractor in the cane-fields of Gujarat was republished almost two decades later in the collection edited by Patel (2008: 212–287), but now with a sub-title altered and changed to 'neo-bondage', although as before neo-bondage and unfree labour were not mentioned.

35 At this conjuncture, therefore, his view (Breman, 1985: 311) was unambiguous: it 'is no longer the case' that permanent farm servants are bonded, 'even when ... indebtedness is a persistent feature of the labour relationship', and although casual workers receive loans and maidservants live in the house of their employers, 'their situation does not constitute an unfree working relationship either'. His view then was categorical and unambiguous: 'To my mind', he asserted, 'it is unsound to deduce from this [the existence of debt] that unfree labour continues in either the same form *or a new one* ... [t]he binding which accompanies this cannot ... be equated with unfree labour' (emphasis added). Similarly, 'I shall regard as unfree only that form of debt-labour which is rooted in non-economic coercion ... *this relationship has nothing to do with the essence of present-day control over agricultural labour*' (emphasis added). What was happening, by contrast, was 'the transition from a traditional agrarian economy to *a free labour market* in the countryside [and]

nothing if not adamant, taking issue with the deproletarianisation approach in a forthright manner. Dismissing the deproletarianisation argument that what was happening amounted to 'replacing free workers with unfree equivalents or by converting the former into the latter', his view then was that 'such reasoning implies that a process of capitalist transformation is in progress in the Indian countryside in which free labour is disappearing to make place for a regime of unfreedom', concluding unequivocally: 'In fact the trend is the reverse'.

Misinterpreting Capitalism

Most of the difficulties and contradictions informing the analyses by Breman of the historical transformations in the production relations of rural Gujarat stem from his problematic interpretation of theory. Criticizing Marx because he 'welcomed capitalism', thinking this would enable 'the dispossessed proletariat' to become free wage-labour, Breman dismisses this view as 'time-bound and Eurocentric', on account that 'the wisdom of hindsight has clarified that capitalism and unfree labour are not inimical at all' an interpretation attributed to Eric Williams.[36] None of these arguments is correct. To begin with, Marx did not regard the replacement of unfree labour by its free equivalent as automatic, seeing it much rather as dependent on the balance of forces in the class struggle. Nor did he argue that unfree labour-power is incompatible with capitalism, much rather the opposite, as shown by the deproletarianisation approach. Equally problematic is the accusation that Marxist theory is Eurocentric, an inaccurate trope emanating from conservative ideology and the 'new' populist postmodernism. Similarly misinterpreted are both the origins of the case made by Williams, and the case itself.[37]

---

the acceleration of a capitalist mode of production in agriculture' (Breman, 1985: 443–4, emphasis added).

36 Breman (2023: 29).

37 Like Patnaik earlier (on which see Brass, 1995), Breman wrongly attributes the capitalism/unfreedom connection to Williams, whereas as the latter himself states (Williams, 1964: 268) this argument was made initially by C.L.R. James (1938). Moreover, the argument itself is not as Breman imagines. The case made by Williams concerned not the acceptability to capitalism of labour-power that is unfree but much rather the opposite: because it had become an obstacle to further development, slavery was destroyed by industrial capitalism. By contrast, deproletarianisation indicates that in certain circumstances industrial capitalism actually depends on the continuation of unfree labour, which is the mirror image of the Williams thesis.

Far more serious in terms of its implication for the case Breman seeks to make about neo-bondage and capitalism is his faulty comprehension of what constitutes a production relation. Throughout the present book, as with previous volumes, workers categorized as unfree are described wrongly as composing a proletariat.[38] This despite the fact that labour which is relationally unfree cannot be regarded as such. What is missed, therefore, is that if labour does not meet the condition of double freedom stipulated by Marx and other political economists, then it also cannot be regarded as belonging to a proletariat, since membership of the latter is determined by adherence to each of these two freedoms, only one of which is – as Breman himself now accepts – fulfilled. In short, as Marxism generally and the deproletarianisation approach in particular have long argued, workers who are unable personally to commodify their own labour-power are not – and cannot be considered as – forming a proletariat.

Ironically, this does not prevent him from eventually recognizing the centrality to the accumulation process of workers that are doubly free, and indeed pointing out that neo-bondage contradicts this stipulation.[39] Notwithstanding his acceptance of 'the classical assumption' that capitalism requires doubly free labour-power – a worker must be free not just from the means of production (land, tools) but also to sell his/her own labour-power – Breman nevertheless insists on classifying as a proletariat those who meet only one of the two

---

[38] That Breman really seems not to understand what a proletariat means is evident from an earlier attempt by him to question the deproletarianisation approach. Then he argued (1993: 301) that '[i]n fieldwork carried out in 1987 in five villages in Haryana, [Brass] found that farmers used debt-bondage to try to prevent landworkers from opting for alternative and better-paid employment. Brass speaks in this respect of capital's de-proletarianization of labour … In my opinion, his interpretation of the situation is debatable on various grounds. The way in which he characterizes the dependency of agricultural labour is, moreover, extremely unfortunate [since] submission to capital engendered by means of a debt relationship causes Brass to deprive present-day farm workers in his research area of their proletarian status'. This demonstrates as clearly as necessary that, because he defines membership of the proletariat only in terms of landlessness, and not also of an untrammelled capacity on the part of a worker to sell his/her own labour-power, Breman fundamentally misunderstands who does – and who does not – belong in that category.

[39] 'To sum up', notes Breman (2023: 162), 'the disappearance of agrestic servitude in most parts of south Asia, in the manner in which it operated in the pre-capitalist past, has not done away with exploitation and oppression of the swelling proletariat. The wage-dependent workforce in the nether echelons of the economy and society is not in a position to themselves determine when, where and for whom to work. That conclusion contradicts the classical assumption that capitalism is marked by a mode of production based on freedom of labour in a double sense: free from the means of production, as well as free as to when, where and to whom to sell their labour power'.

conditions for membership. Seemingly in defiance of this difficulty, Breman concludes somewhat perversely that, because unfreedom continued, the ending of 'agrestic servitude' associated with 'the pre-capitalist past' did not signal also the end of 'exploitation and oppression' of what is termed by him 'the proletariat'.

What Breman appears to misunderstand is that, while there may indeed have been changes in the labour regime (from permanent to casual employment, from locals to migrants, etc.) – all of which were noted by and are consistent with deproletarianisation – there is a crucial aspect which nevertheless endures: the unfree element of the relation, in that its subject remains unable personally to commodify unconditionally his/her own labour-power elsewhere without the consent of the creditor (a landowner, a merchant, a labour contractor) from whom a loan has been taken as long as the debt remains outstanding. This underlines as clearly as need be the element of contradiction at the heart of his case. He states, in effect, that bonded labour had 'withered away', but in fact it continued ('lingered on'), and moreover, in the way ('abrasive domination') that workers were not only treated but succumbed to this ('the muted acceptance by the subordinated workforce'). Its significance is that it undermines the main claim Breman makes about the pre-capitalist *hali* system and how it changed once capitalism established itself. His argument on which the case for neo-bondage rests is that workers no longer accepted earlier forms of 'abrasive domination', yet he unwittingly confirms the opposite: that they did indeed accept – mutedly – just such treatment.[40]

Conclusion

The story Breman wants to tell in this book is twofold: that it was he himself who spotted the link between capitalism, unfree labour, and neo-bondage, and that it was a connection already in place by the late 1970s. Because of this epistemological continuity and analytical consistency on his part, therefore, no change of mind on his part occurred. Notwithstanding the desire to put in place a chronology supporting this claim, it is an agenda hard to sustain.

---

40   'Without detracting from my contention that bondage had withered away', contends Breman (2023: 105, emphasis added), 'it was clear that [in 1986–87] remnants of what had been passed from generation to generation lingered on and could be detected in the manner in which members of the main landowning caste treated their regular workers. *Their abrasive domination appeared to be echoed in the muted acceptance by the subordinated workforce*'.

Breman presents the case about his research as though it was the context that changed in the intervening period between fieldwork in the early 1960s and that in the late 1980s, whereas what has changed is his interpretation of the context, from that expressed initially to the view taken subsequently, not the context itself. Having encountered the deproletarianisation framework in the interim, he decided, not unreasonably, that this offered a better explanation than his own initial one as to what was happening on the ground. He changed his interpretation radically, from free labour as the outcome of capitalist development to neo-bondage, bringing it into line with deproletarianization – a concept he earlier rejected as inapplicable – without, however, indicating that this is what had been done.

The need to change his mind about the unfreedom/capitalism link in this manner is due in part to intellectual shifts in the way bonded labour itself has been interpreted, misinterpreted, and reinterpreted. During the 1960s development decade, unfree production relations were perceived by modernisation theory and the semi-feudal thesis as an obstacle to economic growth, which producers sought to replace with labour-power that was free, was the dominant paradigm when Breman carried out his initial fieldwork and a view from which he did not then dissent. From the 1990s onwards, however, this interpretation was reversed: it was argued that instead of dispensing with bonded labour, and replacing it with free equivalents, an unfree workforce was much rather the preferred relational form – conceptualized by him as neo-bondage – reproduced or sought by agrarian capitalists.

Having argued strongly for the 1960s interpretation, Breman was faced with something of a dilemma: either he could admit it was incorrect and move on, or else maintain that his view had always been consistent with the 1990s interpretation. Since the latter informs his publications from that point onwards, it is this claim which has been examined here. As the most recent book by Breman constitutes a summation of his views both about the changes in the labour regime of Gujarat and in particular the unfreedom/capitalism link there over the past half century, it offers the possibility of assessing the consistency of his interpretation over this period. Because the narrative possesses as its sub-text an attempt to establish an earlier recognition of the acceptability to capitalism of unfree labour, therefore, the question that requires answering necessarily concerns the accuracy of such an account.

PART 3

Beyond Capitalism?

CHAPTER 9

Postmodernism and Development: Misremembering the Peasantry

> This [excavation] site also demonstrates one of the great dangers of archaeology ... I'm talking about folklore.
>
> A warning about false trails, by Indiana Jones to his students, in the film *Raiders of the Lost Ark* (1981), directed by Steven Spielberg.

∴

Introduction: Doing without Development?

This and the following chapter focus on the way postmodern theory has become hegemonic in the social sciences generally, and development studies in particular, a colonization achieved via social history through its claim to have revealed what the grassroots subject really thinks, believes, and does. From this it is but a short step to the view that, because what is found seemingly does not accord with Marxist theory about class, the latter is no longer analytically valid. Henceforth, so this postmodern argument goes, it is non-class identity that directs agency, the object of which is to recuperate selfhoods based not on class but on nationalism and/or ethnicity. Since the latter discourses, together with their forms of political mobilization, derive from the grassroots, such ideology is deemed not just 'authentic' but also benign, not to say progressive. This is an interpretation examined here with reference to the manner in which social history claims to have reinterpreted the peasant voice as a result of examining it through the postmodern lens. It is the adequacy, or otherwise, of this sort of claim that is considered here.

The author of a recent book about the peasantry, Patrick Joyce, comes with the kind of theoretical and political baggage that has implications for the case he seeks to make about its subject.[1] His area of study hitherto has been the social and cultural history of the urban working class in England over the

---

1   Joyce (2024).

nineteenth and early twentieth centuries, to which he has applied the theoretical apparatus of postmodernism.[2] The result is, to put it mildly, extremely peculiar, since we are told in somewhat peremptory fashion that henceforth all the usual social scientific categories associated with any analysis of the urban workforce are redundant, not to say forbidden. Among the main concepts dismissed in this manner are class, modernity, the Enlightenment, grand narratives, social determination, totality, and ultimately reality itself, as embodied in the presence of an objective material existence beyond and apart from the realm of its many discursive interpretations/descriptions.[3] In their journey towards the historical dustbin these concepts are to be accompanied by two kinds of consciousness: that of class, and that which is false.[4] It comes as no surprise that on this very same journey to historical and political oblivion is Marxism, the *bête noire* of postmodernists everywhere.[5]

---

2   For the background to this, see Croll (2002). Lest it be thought that the postmodern framework of Joyce meets with the universal approval of fellow historians, one such (Perkin, 1996: 10) has criticized his views strongly, observing that 'In the face of such eye-witness evidence it is difficult to agree with some "post-modern" historians who reject the very existence of class in the nineteenth century or dismiss it as only one factor among many, notably populism, gender and patriotism … Sometimes the approach verges on obscurantism, as when "The consciousness of a class need not, and has not been, the consciousness of class," or working-class self-expression is transmogrified into "labourist populism"'.

3   These proscriptions are outlined in Joyce (1982; 1994; 1998; 2010). Downplaying not just class but also struggle based on this, he (Joyce, 1991: 64, 66) maintains 'how necessary it is to avoid anachronistic understandings of class as primarily "economic" and conflictual in character', because of the impact of 'a radical, reforming bourgeoisie, bent on the politics of class union'. This benign view has been projected forward to the present (Joyce, 1998: 225): 'To the extent that a belief in "objectivity" has been tied to the fate of the nation-state, with a decline of a powerful and coherent Britain after 1945, this belief has clearly taken a knocking, and there does seem to be a connection between the onset of a "postmodernist" relativism and the general de-centring of former concentrations of power'. Much rather the opposite is the case, however, in that postmodernism has reinforced, not decentred political power, since its message of replacing class identity with non-class equivalents, and consecrating the latter as innate/empowering, has played directly into the hands of conservatives and neoliberals, each of whom shares a similar ideology concerning identity and its durability.

4   In the words of Joyce (1991: 9) class consciousness 'has indeed an antiquated ring to it, one redolent of a time, not so long ago, when class was seen in terms of … what were in fact hopelessly idealised categories such as "revolutionary" or "labour" consciousness, notions emanating from … earlier Marxism'. To this is added the following: 'The reason why such notions of consciousness have become superannuated is above all the effect of the new historical interest in theories of language and ideology … [i]n the disaggregation of "class", deconstruction has taken this route of language, ideology and identity. This is the route of this [1991] book'.

5   On the epistemological incompatibility between Marxism and postmodernism, together with the reasons for this political and ideological antagonism, see Brass (2000; 2021b). Taking

In their place historians and other social scientists are required to insert categories like alterity, difference, ambivalence, and contingency. This is because, according to postmodern theory, outside of language there is nothing, no 'out there' exists, only stories about 'out there'; that is to say, no objective material reality, only competing accounts of this. In this way, history is reduced to a contesting series of subjectivities, between which it is not possible to decide. A consequence of this aporia is that politics and political mobilization – especially that informed by Marxist theory about the desirability/feasibility of a socialist transition – become virtually impossible, since they have in effect been deprived of a material referent.[6] Socio-economic processes and relational forms such as exploitation and coercion vanish because conceptually they have ceased to exist, and thus cannot be invoked as issues either to be opposed politically or on which to build a programme of action designed to bring about change.

Unlike Marxism, which regards populism as the anti-capitalism of the right, Joyce sees it as a benign – not to say empowering and progressive – alternative political identity espoused by labour, an 'authentic' expression of the plebeian voice-from-below, claiming that '[w]hen we come to the matter of definitions, it will be seen that if class has a rival it is perhaps that of "populism", of "the people"'.[7] Contrary to what is argued by Joyce, however, currently and historically populism has been – and is – itself part of the class struggle, waged from above by capital with the object of generating or reproducing false consciousness among its workforce. This it does by deploying the age-old and well-tried strategy of divide-and-rule, encouraging worker perception of

---

issue with Marxism by advancing the postmodern claim about no class, no worker, and no struggle, Joyce (1991: 3) argues that 'the term "proletarian" applies with only a good deal of qualification [resulting in] a distinctly tenuous "working class". This questioning of the idea of the proletarian is furthered by new considerations of the relationship between labour and capital. Instead of an overmastering, trans-historical tendency towards conflict – along classical Marxist lines – what is evident is the inter-dependence of capital and labour ... The vested interest workers and employers have in co-operation is at least as great as any tendency towards conflict ... indeed, socio-economic class position or situation emerges as so fractured and ambiguous that the very notion of class may itself be questioned'.

6 According to Joyce (1991: 208–209), '[t]here is no overarching coherence evident in either the polity, the economy or the social system ... there is no underlying structure to which they can be referred as expressions or effects ... [t]hus with the notion of social totality goes the notion of social determination ... [t]he certainty of a materialist link to the social is likewise broken ... [g]one too are the grand narratives that historicized the notion of social totality'.

7 See Joyce (1991: 5–6), whose interpretation of British working class conservatism is not in fact new, a similar case having been advanced before – and more critically – by McKenzie and Silver (1968) and Roberts (1979), among others.

labour market competition as an effect of non-class identity. To postmodern discourse emphasizing the cultural identity of the migrant-as-'other'-nationality/ethnicity, therefore, conservatives and the far right counterpose an argument similarly privileging cultural identity, only this time the nationality or ethnicity of the non-migrant worker.[8]

Now, it seems, Joyce has taken upon himself the evangelical task of spreading this same postmodern gospel to those who study peasants and rural society, little knowing that not only has this already been done (subaltern studies, multitudes) but also been strongly criticized. Accordingly, he dusts down and redeploys unwittingly all the old tropes informing 'new' populist postmodern arguments about the peasantry, as though these have yet to be discovered – let alone debated – by those who conduct research on peasants. It soon becomes evident that, in terms of the sources/methods/theory needed for an analysis of this non-urban topic, Joyce is clearly outside his comfort zone. For this reason, among others, the book is yet another instance of a case strongly made in the first half unravelling as the narrative proceeds.

This chapter consists of two sections, the first of which examines the methodology applied by Joyce in his account of the peasantry. The adequacy or otherwise of the theory used by him to interpret peasant economy, culture, and politics is considered in the second section.

I

Methodology

The postmodern rootedness of the approach is difficult to disguise, not least in the way the self is positioned front-and-centre in the story about peasants. This tendency to write oneself into history cannot but give rise in some instances to what is unmistakably a me-me-me narrative.[9] While it is certainly legitimate to self-cite where one is engaged in debate with those holding different interpretations, or writing a straightforward autobiography, to do so in the manner

---

8 Nowadays evidence suggests that in these circumstances a radical politics remains on the agenda, but with the difference that working class support can be transferred instead to right-wing populist movements offering empowerment on the basis of nationalism and/or ethnicity. For more on this, see Brass (2017b: Chapter 19).
9 In the course of an interview published in *The Guardian* (London) on 3rd February 2024, Joyce states that 'I was an insider to the peasant world. I knew something about it'. Not only is this moot, claiming an insider status which peasants almost never concede to non-peasants (and even some other peasants), but it reveals the extent to which the methodology on which the narrative is based is rooted in the history of his own kin.

Joyce does comes perilously close to a vanity publication.[10] This is because methodologically grounding an analysis of a particular socio-economic category mainly on the experience of one's own family history makes it ontologically difficult to avoid privileging subjectivity (what happened to my kinfolk and why) over objectivity (what happened to everyone else, and why).[11] Were it the case that those in the latter category – peasants subsumed under the general rubric of 'everyone else' – feature in the book in a theoretically rigorous and epistemologically less one-sided manner, then even the privileging of subjectivity need not of itself be a huge drawback: that this is not the case merely compounds the problem.

Unlike the 'experts' Joyce derides, for the most part anthropologists who have undertaken long-term fieldwork in rural contexts, he appears not to have conducted the equivalent sort of research outside his own kin group, finding out what rural people – other than those to whom he is related – think, say, and do. As in the case of others who similarly privilege direct engagement with the rural grassroots as the only valid methodology, yet lack this kind of research experience, Joyce relies overly on what other published sources record about peasants, but himself has nothing in terms of fieldwork with which to compare what is learnt from his kin.[12]

Joyce is critical of ethnography because in his view peasants are heard only as a mediated voice.[13] Yet he does not consider in any detail the extent to which the peasant voice is itself mediated, let alone its effect, and projects ideology and concepts of those – landlords, merchants – who are not themselves peasants.[14] As presented by him, therefore, the notion of a mediated voice

---

10 In a social science text any methodology built around linkages to self and kindred elements does indeed constitute a fine line that one has to tread with care; when not done so it is all too easy to generalize what are in fact nothing more than specific experiences, transforming the latter into nomothetic instances. An example of the way in which self and other can be successfully integrated in a single ethnographic narrative is Hart (2022); reasons why the combination works are set out in Brass (2023c) and also in Chapter 7 above.

11 This is not seen by him as a difficulty, much rather the opposite: he describes himself (Joyce, 2024: XV) as 'the London-born child of Irish rural immigrant parents ... a sort of relict of what we have lost', adding that 'I write this book [as] a homage to my own'. References by Joyce to the experiences and background of his own kinsfolk surface constantly throughout the book (see, for example, Joyce, 2024: 5ff., 17ff., 25, 60ff., 95, 116–17, 119ff., 132–33, 138, 171, 203, 207ff., 321ff.), so much so that they can be said to be a major aspect of the narrative.

12 A similar methodological difficulty confronts van der Linden (2023), whose global labour history framework privileges fieldwork experience that he himself does not have.

13 Joyce (2024: XIV).

14 Even a source cited by Joyce acknowledges this, noting (Thomas and Znaniecki, 1927: 302): 'More recently an intense aesthetic movement has manifested itself among

assumes – wrongly – that underneath all the accretions lies an 'authentic' peasant voice, unsullied by anything deemed to be a non-peasant influence.[15] For someone who dismisses even those who conduct long-term village-level research as purveyors merely of a 'mediated voice' – an inauthentic discourse – Joyce puts a lot of faith in published accounts of peasant culture, not least by those who write about 'what peasants believe' and why.[16]

A related methodological issue concerns the range of peasant societies covered by the book. It is necessary to ask, therefore, how is it possible to generalize about the meaning, the social history, culture, and economic role of *the* peasantry in an analysis that draws its main examples solely from European history (Italy, Ireland, Poland), making scant reference to non-European contexts.[17] Equally problematic in terms of methodology are statistics purporting to show the decline of the peasantry ('people employed in agriculture'), presented without any indication of how this category is defined, an exercise undermined by the many different and incompatible criteria that go into the construction of these data. Observing that peasant numbers in Spain halved over the 1950–70 period, no mention is made by Joyce of how such a category is defined in this particular context, despite the caution invoked by other researchers who have emphasized the difficulty identifying who is – and who is not – a peasant where Spanish landownership records are concerned.[18] Subsequently, however, this is contradicted: 'Perhaps the "death" of the European peasantry has been exaggerated'.[19]

---

peasants, particularly along literary lines, and while this is developed upon the traditional background it tends increasingly to come under the influence of the models presented by the upper classes'. As the narrative proceeds, Joyce qualifies his view, noting that (Joyce, 2024: 117) peasant 'lives were literally more raw and less mediated than ours', without explaining how and why this is so.

15   Joyce (2024: 35). As will be seen below, this is no different from what Chayanov says about peasant economy, and just as mistaken.
16   This becomes increasingly evident in Chapters 4 and 5 of his book, which make extensive use of stories told by peasants about life-cycle beliefs, including the efficacy of magic, fairies, and vampires.
17   There are occasional references to peasant society outside Europe, but these are only in passing, without an attempt to examine such contexts in any detail.
18   See Joyce (2024: 12, 13ff.). On this methodological difficulty, see for example Malefakis (1970).
19   Joyce (2024: 14).

Sources

Unsurprisingly, therefore, the publications about the peasantry on which his case is based consist for the most part of journals and books either by postmodernists or by agrarian populists, all of whom are cited endorsingly.[20] In keeping with this the few references to *The Journal of Peasant Studies* (JPS) are confined largely to post-2008, at which point it changed into a platform for agrarian populism.[21] By contrast, what Marxism has to say about the peasantry, as represented by its many contributions to these debates, both in the pre-2009 JPS and elsewhere, is notably absent. The views of Marx and Engels are deemed to be of no consequence, those of Lenin, Trotsky, and Kautsky are passed over in silence, while the sole reference to Rosa Luxemburg is to her assassination, not her ideas.[22] Not far into the narrative one encounters, inevitably, the conservative trope equating Communism with Fascism.[23]

One deleterious effect of problematic methods and unreferenced sources is the number of empirically questionable judgements scattered throughout the book that are accepted at face value. Thus, for example, peasant networks described by Joyce as reciprocal and equal exchanges, whilst based notionally

---

20 Positive, not to say enthusiastic, references are made by Joyce to postmodernists such as Michel Foucault (Joyce, 2024: 118, 346 note 12) and Jean Baudrillard (Joyce, 2024: 362 note 10), and also to agrarian populists like A.V. Chayanov (Joyce, 2024: 22–23, 335 note 1), Teodor Shanin (Joyce, 2024: 43, 333 note 9, 335 note 1, 337 notes 3 and 5), Jim Handy (Joyce, 2024: 337 note 18), Marc Edelman (Joyce, 2024: 338 note 7), Jan Dowe van der Ploeg (Joyce, 2024: 338 note 7), Henry Bernstein (Joyce, 2024: 339 note 19), and James C. Scott (Joyce, 2024: 355 note 27). As effusive are the frequent mentions of John Berger, whose ideas about peasants loom over the entire case made by Joyce (2024: X, XI, XIV, 37–38, 63, 98, 99, 115, 119, 163, 164, 231–32, 238, 245, 298–99, 301).

21 Incorrectly described (Joyce, 2024: 337 note 4) as having a 'perspective [that] is a strongly political economy one', JPS contributions to on-going debate about the peasantry are limited to what is said in a contentious account (Joyce, 2024: 339 note 19) by an agrarian populist.

22 For this reference to Rosa Luxemburg see Joyce (2024: 315–16). That socialism and leftism generally have no place in the narrative is evident from observations such as (Joyce, 2024: 12) 'the strange temporality of Communist governance' where peasants had their lives transformed through collectivization and 'enforced co-operative methods of farming' which was 'in most places a degradation', while a Polish ethnographer is commended (Joyce, 2024: 123) for having 'defended [peasant] interests against the excesses of Communist rule'.

23 The claim (Joyce, 2024: 16) that '[i]n authoritarian countries, Fascist and Communist both, people were long kept on the land uneconomically' is contentious on so many different levels, not least because the *lebensraum* policy of the Nazis was based on conquest and resettlement in order to provide food for the Reich, so its objective cannot be regarded simply as an uneconomic one.

on such discourses, often hide what are capitalist production relations, operating between and within peasant households.[24] As problematic is the positive way in which some of the sources are described: that *Into Their Labours* by John Berger is a 'remarkable work on late twentieth century peasant experience' is open to dispute, as is 'the great social historian' label applied to Eric Hobsbawm.[25]

Similarly moot are uncritical references both to Teodor Shanin as having 'defined the academic study of peasants for decades' and to the JPS which 'continues to do this'.[26] Equally doubtful is the benign portrayal of the Sicilian mafia as siding with peasants in 'solidarity against outsiders', rather than peasants who sided with landlords against other peasants.[27] In keeping with the earlier postmodern approach, there are references to the body of the peasant, including the claim that it 'was a burden to him and a deterrent in the marriage market'; this despite the fact that the deterrent was economic – peasants invariably did not earn enough to marry – and nothing to do with the body *per se*, an explanation accepted subsequently.[28]

Stories

In the light of these sorts of methodological difficulty, it comes as no surprise to see this confusion play out with regard to what Joyce considers to be a useful source for the study of peasants: folklore.[29] Lamenting the failure of 'experts' to

---

[24] Joyce (2024: 44). For instances in rural India and Latin America of discourse about reciprocity and equal exchanges, involving peasant households and/or kin groups, that in reality mask employment arrangements between capitalists and labour, see Brass (1999).

[25] For these descriptions of Berger and Hobsbawm, see Joyce (2024: x). Critiques of the way in which each of them interprets peasants are found in Brass (2018a: Chapters 9 and 10).

[26] Joyce (2024: 43). On Shanin see the critique by Littlejohn (1973a, 1973b,1977), and on the post-2008 JPS see that by Brass (2023a).

[27] Joyce (2024: 47) appears to reproduce the the same idealized image of rural banditry as Hobsbawm (1969), an interpretation challenged by Blok (1988) who pointed out that the Sicilian mafia were enforcers of landlord power in the countryside. A similar difficulty is evident in the benign way Joyce (2024: 52) describes debt bondage; not as a coercive production relation, but as a form of reciprocity whereby '[i]n return for their obligations to the landlord, peasants were entitled to protection and held certain rights in respect to the land they cultivated'.

[28] On the body of the peasant, see Joyce (2024: 19–20, 24, 62, 117).

[29] 'The original folklorists were children of their time', accepts Joyce (2024: 123–24), 'like all of us, dressing the peasant up in clothes of their own choosing … [r]eligion, like patriotism, was a good thing. What was "modern" corrupted the soul of the peasant', adding (Joyce, 2024: 125, original emphasis) with regard to 'the stories of peasants and what they

ask peasants about smoke, shit, and sex, Joyce neglects to indicate why should these topics should be the object of enquiry, which suggests that methodologically he is simply following in the footsteps of the antiquarian.[30] For the latter, such issues are of interest in order to highlight the 'quaintness' of peasant life, along the lines of 'look how they do these same kinds things as us, but differently'.[31] Beginning, deceptively, by querying the usefulness of folklore, Joyce concedes that even if asked, such questions would not be answered, which leaves unposed and thus unanswered what then is the point of this 'absent' data. Nevertheless, he quickly abandons caution and moves on to recommend folklore as a source, asserting that we should be grateful for what it provides in terms of information, without raising the issue of how to evaluate this, by interrogating the data/narratives/stories which remain.[32]

Having elaborated unquestioningly a series of peasant 'beliefs' (in Chapters 4 and 5 of his book), Joyce at last gets round to the question of 'how do we know', which is ironic in view of the fact that much of the preceding narrative is spent on casting doubt on what has been written about peasants by planners, development theorists, and anthropologists (= 'experts').[33] He then asserts that knowledge of this kind – about peasants – is obtained as a result of 'the stories peasants tell', which merely returns one methodologically to the above point – mediated voices are suspect in his eyes, yet it is on these very same sources that Joyce relies for these stories. This contradiction notwithstanding, his view is

---

tell us about peasant beliefs [that] like folklore, and indeed, for me, *folk* anything, "belief" is a tricky subject'. The latter caveat is indeed ironic, given his subsequent endorsement of folklore as a source of information on the beliefs and much else held by peasant society.

30 On this Joyce (2024: 79) notes: 'Experts on peasants, and folklore ... upon whom so much of our basic historical knowledge depends, did not, with few exceptions, write of smoke, nor of dirt, defecation, delousing or sex. The list of omissions is very long. If these experts had asked peasants about these things, they would not have been told anything anyway ... What was not (and is not) told to investigators would fill many more volumes than what was, but we depend on this "what was" bit, and in truth should be grateful for that'.

31 'The Folklore Commission people were forever interested in "survivals", antiquities, old wisdom', remarks Joyce (2024: 121), adding that 'behind this recording of the present was the feeling of the folklorists that the old was better, and must be recorded before being lost ... they did not like what the advance of modern life brought with it, not unreasonably in many cases' Again the irony is unmissable, since this description of the folklorist is one that applies to Joyce himself, in that he, too, writes about peasants before they are lost (hence the centrality of remembering), and who – like them – disapproves of what modernity brings.

32 Significantly, earlier Joyce (1987: 3) also commended folklore as a source of information about urban industrial society, noting that '[t]here are other currents that are also of value, for instance work based on folklore studies'.

33 Joyce (2024: 119).

that '[f]olklorists have changed', and because of this have influenced what he himself writes about peasant stories.[34]

Such methodological inconsistency is justified in turn by the invocation of 'reliable sources' voiced by those deemed 'reliable persons', neither of which terms are defined.[35] Instead, the veracity of peasant belief is declared rational on the grounds that for its subject it 'works', which reproduces the postmodern trope that privileges anything said by a 'from below' voice, regardless of its content. This licences the idealization of myth over history, and with it the claim there is no difference between them, which is incorrect: it also replays the old debate on this issue between Sartre and Lévi-Strauss.[36] Stories are heard and passed on – changed – by those who have no (and cannot have) experience of what the source (whether 'reliable' or unreliable) was, to what it referred, plus why and how it continued to circulate thereafter. Such accounts are not – and cannot be – deemed 'rational' in the wider and more accurate sense of describing episodes/experiences beyond the immediate vicinity of the self. By privileging such discourse, Joyce reproduces one of the central tropes of postmodernism: namely, that there is no such thing as *false* consciousness, an ideology that does not reflect the true interest of class, preferring rather to claim that *all* consciousness is true. Thus what peasants believe to be true is *ipso facto* true, and as valid as any other kind of belief in terms of a reality it purports to describe. As such, the inference according to Joyce appears to be that it cannot be challenged.[37]

---

34 Joyce (2024: 121).
35 'These [folkloric] truths', contends Joyce (2024: 127), 'have to be gathered not only from reliable sources but also told by reliable persons ... All this "belief" is counter to our versions of rationality. however, what must be emphasized is that peasant cosmogonies are entirely "rational", which is to say, they are coherent, grounded in experience. They are clear-eyed means of getting things done using methods that have worked'.
36 Asked by Eribon (Lévi-Strauss and Eribon, 1991: 125) about this debate with Sartre, Lévi-Strauss replied: 'When Marxists or Neo-Marxists attack me for not knowing history, I answer, You are the ones who don't know it or are turning your backs on it, since in place of history you set up grand developmental laws that exist only in your minds'. The answer given by Lévi-Strauss to the next question posed by Eribon, 'You reject the idea of "historical laws"?', is revealing: 'There are so many variables, so many parameters, that perhaps only a divine understanding could know or does know what happens and what is going to happen for all eternity'. In other words, deploying against the historical materialism of Marxist theory the all-too-familiar aporia of postmodernism.
37 Hence a belief in fairies or vampires is as valid as a belief that if a person jumps out of a skyscraper window s/he will fall to his/her death. One of these beliefs is true, the other is not.

II

Theory

Where theory is concerned, every single populist trope about the peasantry is deployed, among them claims about opposition to development/modernity/progress and/or absence of class differentiation and capitalism in the countryside. Typical, perhaps, is the view that 'the great victims of modernity and progress have been peasants' who 'were, after all, right to distrust progress'.[38] Development itself is condemned as a misbegotten attempt at the 'social engineering' of peasants, an indictment aimed not just at the Enlightenment but also at Marx and Engels, '[d]amned by the self-evidently true dicta of political economy [which maintained] that humanity existed so that it could be improved'.[39] That the political voice one is hearing is that of Joyce himself is evident: hence the impression conveyed throughout is that denials notwithstanding – 'we cannot go back and be peasants', and 'I did not write this book to make peasants tutors to the present' – he is indeed advocating just this; learning from an idealised image of the peasant as part of nature.[40]

In what is yet another anti-development/anti-modern trope deployed by the 'new' populist postmodernism, Joyce also declares that 'peasants, being part of "tradition", could not partake of the history that was the hallmark of modernity' since '[t]hey were outside history'.[41] Symptomatically, therefore, when posing the question about peasant-as-victim ('How has this come to be? What is behind the curse?'), his answer is not capitalism but time. Similarly mistaken is his view that peasants are excluded from modernity: they do indeed participate in the latter, but either as small capitalists or as workers. The viability of the middle peasant is eroded by the economic forces of which s/he is a part, and thus not by time but by capital, as Lenin and others long ago pointed out. A consequence of an unwillingness to engage with Marxist theory about development, modernity, and class, manifests itself centrally in problems with defining who is a peasant, and why.

Definitions

From the start, attempts by Joyce to define the peasantry generate a certain amount of confusion and contradiction. Moving quickly from terms like 'small

---

38 Joyce (2024: XII).
39 Joyce (2024: 36).
40 Joyce (2024: 115).
41 Joyce (2024: 37).

farmer-cum-peasant', and 'hobby farmer', he opts early on for a rudimentary description: 'A peasant is a country person, a person of the land. That is all the word in its original innocence means'.[42] About the logic informing smallholding agriculture, however, Joyce is adamant: 'The essential point is that land ... is understood to be be a social rather than an economic activity', since peasant landholding is in his view 'a family economy and not a business ... the family economy denotes a way of life'.[43] The incompatibility between the latter model and capitalist production is underlined thus: 'We need to be aware that peasant societies are societies of the gift, not commodity, like our societies', buying into anthropological theory about reciprocity and gift exchange without interrogating these concepts.[44] In arguing that 'peasants are the most dominated class [because] they are poorer, less educated, more subject, from oral cultures and so on', he reproduces another populist trope, defining peasants as a class.[45] As Marxists have pointed out *ad nauseam* over the years, peasants are not a class but a category that is itself transected by class.

Although the early confidence as to what the definition is soon begins to break down, Joyce nevertheless insists on retaining what he sees as the core elements of the peasantry: culture as its defining aspect, the family as its *raison d'etre*, and subsistence as its economic object.[46] For this reason, he is adamant in wanting to exclude better-off producers from the peasant category, in effect reducing the latter to subsistence cultivating smallholders.[47] Maintaining inaccurately that no peasant is interested in the profitability of his agricultural enterprise, Joyce cites with approval the Russian neo-populist economist A.V. Chayanov in support of the twofold claim that, as peasant economy is simply about family subsistence, the rich peasant capitalist or kulak cannot be anything other than a mythical figure.[48] Seemingly unaware of the debate

---

42   For these definitions, see Joyce (2024: 6, 17, 21).
43   Joyce (2024: 23, 24).
44   Joyce (2024: 98).
45   Joyce (2024: 35).
46   'The distinctions between peasants', he accepts (Joyce, 2024: 42–43), 'are as various and obscure to us as are the criteria of measurement they employed in defining the land that meant everything to them ... the church of peasants is clearly a pretty various one ... [t]here is difference among peasants but also fluidity and change'.
47   See Joyce (2024: 11). This exclusionary approach is clear from, for example, the view that (Joyce, 2024: XII) 'peasants come from a world that in essence is not capitalist, although they have coexisted with capitalism for centuries'.
48   Hence the following (Joyce, 2024: 14, 22): 'That which creates maximum reward from what is available to the farmer is surely "efficient", to use a word that does not come out of peasant lexicons', and 'Chayanov believed that the logic of peasant production was not conducive to the production of a surplus ... they produced what was needed to survive.

surrounding Chayanovian theory about the peasantry, he simply asserts that Chayanov 'was right', ignoring thereby the many criticisms of this approach.⁴⁹

Not the least important of these objections concerns its closed economic model, whereby the peasant family farm is conceptualized as an hermitic institutional form, unaffected by what occurs outside of itself. This is because the family labour farm was viewed by Chayanov as governed by an endogenous dynamic, in contrast to Marxism which places its reproduction in the wider socio-economic context, and thus subject to an exogenous dynamic over which it has little control. Unaware of the objections to the Chayanovian model, Joyce invokes two interrelated tropes deployed by populists concerning the undifferentiated nature of the peasantry, each of which is wrong. The first is that Chayanov did not defend kulaks, which he did, simply by merging them analytically within the peasantry as a whole, a ploy that made them invisible epistemologically and politically. Second, that the kulak is a non-existent category, which is also wrong, as sources writing at the time and since all testify.⁵⁰

Economy

Having endorsed Chayanovian theory about the peasantry, dismissed the concept 'kulak' as non-existent, and stressed the non-economic function of the family farm, Joyce then contradicts all these views.⁵¹ Not only does he argue that '[n]o class has been or is more economically conscious than the peasants', therefore, but he accepts that 'occupation, consumption, production and the family rarely lined up in the tidy way that Chayanov described', and also that 'more wealthy peasants may also be landowners ... hirers of labour, etc., and in these capacities share interests with completely different groups'. As the narrative proceeds, recognition of peasant differentiation becomes more

---

He was right [but was shot] as a supposed defender of the "kulak" (the largely imagined rich, capitalist peasant)'.

49 That the Chayanovian model of peasant economy, together with the separate components of its conceptual framework (family labour farm, consumer/producer balance, etc.), are the subject of much dispute would be clear to anyone perusing the many entries dealing with these issues in the two published indexes of the JPS, covering the period 1973 to 2004.

50 Earlier sources confirming both the presence of and village-level opposition to kulaks include non-Marxists (Mackenzie Wallace, 1877; Drage, 1904; Stepniak, 1905) as well as Marxists (Lenin, 1964a; Trotsky, 1962, 1967). Later ones making the same case include Brass (2021a,2022d).

51 Joyce (2024: 23, 63).

pronounced. Hence the view that although '[p]easant history is one of want [and includes the capacity to] sometimes prosper ... when they got what they wanted they did not stop being peasants', in effect acknowledging that among their ranks were rich peasants, which did not stop them from being peasants.[52] Ironically, this is precisely what Marxism says about peasant differentiation as an effect of the accumulation process.

Reversing what is said earlier, regarding both the non-economic object of petty commodity production and its non-capitalist nature, therefore, Joyce subsequently and unwittingly adopts the tripartite categorization used by Lenin, dividing peasants into rich, middle, and poor peasant strata.[53] This recognition notwithstanding, Joyce then doubles down and reinstates his original endorsement of Chayanov, insisting that 'the strict, analytic definition is useful, and this is so because what it highlights is the family and the family economy'. This idealization of the Chayanovian peasant family as an homogenous socio-economic unit ignores the fact that class transects the kin group, both as regards production relation and division of labour (working for the household head, both on-farm and off-farm) and property relation (inheritance of holding enables the household head to exercise control over kin group).

Despite acknowledging that peasants work off-farm as well as on-farm, combining seasonality with division of labour, not addressed are two crucial issues, one materialist, the other ideological.[54] The first concerns what sort of on-farm/off-farm work combination is involved, why, how such decisions are reached, and by whom; the second entails the disjuncture between peasant farming and perceptions by its kin components of peasant selfhood, regardless

---

52   Joyce (2024: 38).
53   'Peasants are not strangers to capitalism', notes Joyce (2024: 44), adding that: 'There is much to be gained by producing for the capitalist market ... And of course small peasants become big peasants, with more land, and some of them (a few historically) capitalist farmers ... There were in most communal settings almost always rich, middling and poor peasants'. Much the same is encountered (Joyce, 2024: 49–50) further on, repeating almost exactly not just the characterization of Lenin – albeit unmentioned as such – but also his account of how and why class differentiation takes place: 'Below the relatively few well-off peasants, the middling peasants had enough land to survive and reproduce, prospering at times. The smaller landholders sometimes had enough to enable survival, and, when not enough, work outside the farm was needed as a supplement. Then there were the completely landless'. Even photographic images are described similarly (Joyce, 2024: 83, 84), in terms of 'middling' and 'poor' peasants, as well as 'a better-off peasant's house', while later references (Joyce, 2024: 122, 123) include 'poor peasant family'.
54   This difficulty for the attempt by Joyce (2024: 25) to define petty commodity production in essentialist terms is hinted at in the observation that '[i]t is helpful to think of peasanthood as denoting a condition and not a fixed thing'.

of the extent to which income from the sale by them of labour-power exceeds that gained from the sale of the product of labour.⁵⁵ Notwithstanding the latter, such elements continue to see themselves as peasants, which decouples the pristine family farm concept structuring the Chayanovian model from the ideology of peasant selfhood.

Politics

Turning to the question of politics, and in keeping with his antagonism to Marxism, Joyce cites the description of potatoes in a sack as evidence that Marx himself said peasants were incapable of agency.⁵⁶ However, what Marx was referencing in this comment about rural France was not an inability to undertake agency but much rather the politically different kinds of mobilization involved, as is made clear in the text concerned.⁵⁷ In another criticism aimed at Marxist politics, Joyce cites with approval a statement by Norman Davies that Polish peasants '"were the bearers of a separate civilization, as distinct and as ancient as that of their noble masters", which made for "an ultra-conservative culture" [and] contemporary reformers and modern social theorists have ignored these factors at their peril'.⁵⁸ However, this ignores that Marxists have made the very same point as Davies about peasant conservatism, long before and long after

---

55   In Poland (Joyce, 2024: 122), therefore, 'miners, factory workers and technicians' have 'affinities with peasant existence' since 'in this area, where mining, industry and farming have for approaching two hundred years all been mixed together, it is difficult to know where "peasant" stops and starts. Full-time peasants lived in the same village as industrial workers, who themselves were often "five o'clock farmers", regarding themselves as at one with peasants, "villagers" all of them'.

56   As expressed thus (Joyce, 2024: 44): 'Despite living in separate and sometimes isolated dwellings, families were integrated into communities, and rich networks of exchanges and mutual assistance were customary. These people were not Marx's potatoes in a sack, incapable of joint action'.

57   'But let there be no misunderstanding', comments Marx (1979: 188), 'The Bonaparte dynasty represents not the revolutionary, but the conservative peasant: not the peasant that strikes out beyond the condition of his social existence, the smallholding, but rather the peasant who wants to consolidate this holding; not the country folk who, linked up with the towns, want to overthrow the old order through their own energies, but on the contrary those who, in stupefied seclusion within this old order, want to see themselves and their smallholding saved and favoured by the ghost of empire. It represents not the enlightenment, but the superstition of the peasant; not his judgment, but his prejudice; not his future, but his past; not his modern Cévennes, but his modern Vendée'.

58   Joyce (2024: 58).

he did.⁵⁹ The counter-revolutionary role of peasants was central to Leninist critiques of agrarian populism, a view he shared with Trotsky. Current Marxists make a similar point with regard to the 'new' populist postmodernism, so this issue about peasants and the kind of politics linked to the defence of private property has not been ignored by Marxists, much rather the opposite.

Ironically, those who *have* ignored it are commentators who think like Joyce himself – the exponents of the 'new' populist postmodernism who mistakenly regard as progressive and empowering any/all forms of 'from below' discourse and rural agency, simply because these are grassroots phenomena. In short, these very same forms of 'ultra-conservative culture'. Furthermore, this is the reason for the positive view of peasants, seen as being in the vanguard of the struggle against colonialism in Latin America and India, where they were not just perceived as heroic resisters against foreign oppression but also (and therefore) part of a smallholding agriculture cast as the basis of an authentically nationalist alternative to colonialism.⁶⁰

Finally, in what is possibly the most bizarre argument encountered in the book, Joyce observes that the increase in knowledge about peasants after Shanin produced his edited volume on the subject, published in books and journals that have appeared since then, is almost too difficult with which to get to grips, a surprisingly anti-intellectual observation.⁶¹ If one is going to

---

59   This absence is in keeping with political arguments and/or sources that are either missing or unsupported by bibliographic references. Lenin, of course, is not referenced, but there are others, Marxist and non-Marxist alike, that ought to have been mentioned. Thus, for example, a reference (Joyce, 2024: 34–35) to peasant silence as a form of defence could have been linked to the hidden transcripts argument of James Scott, whose agrarian populist 'weapons of the weak' approach is indeed cited, but later on (Joyce, 2024: 204). That '[h]istorically peasants do not generally speak' (35) has echoes of the agrarian populist subaltern studies framework (Chaturvedi, 2012), while a paragraph (Joyce, 2024: 37) about how Johann Gottfried Herder connected the peasant to the nation contains no mention of where this argument can be found. This information is in fact provided when the same argument was made elsewhere (Brass, 2014b: 268ff., 286ff.), but – again – with a politically different interpretation; whereas the latter – a Marxist view – is critical of the Herder view, Joyce by contrast is supportive.

60   As is outlined elsewhere (Brass, 2000: Chapter 1), this sort of peasant essentialist ideology is encountered in the *indigenista* discourse of José Carlos Mariátegui, Hildebrando Castro Pozo, Haya de la Torre and Fausto Reinaga in Peru and Bolivia, and also in the populist nationalism of Ranga, Deva and Lohia in India. Common to them all was the view that, beneath the alien imposition of a foreign colonial state, an indigenous peasant culture and institutions – embodying what they regarded as an authentically national identity – survived intact, forming thereby a socio-economic blueprint for the future.

61   'Since Shanin's time in the 1970s', writes Joyce (2024: 43), 'there has been a great increase in knowledge about peasants, almost all of it in academic books and journals, the products of which are for most people either too expensive to access or too specialist and

write about peasants in the manner Joyce seeks to do, then getting to grips with these sources and debates linked to them is a *sine qua non* of the task in hand. Similarly odd is the argument that this literature demonstrates that peasants change and do not possess an essentialized identity, not least because this is the paradigm that he himself seeks to challenge. As strange is that this statement about what the literature contains is accompanied by an endnote referencing in support two publications – Edelman, van der Ploeg – which argue precisely the opposite: that an essentialist, undifferentiated category of peasant does indeed exist, unsurprisingly so as they – like Joyce – are both agrarian populists.[62]

Conclusion

Endeavouring to find what is thought to be an 'authentic' rural voice-from-below like the one he claims to have found among urban industrial workers in Victorian England, Joyce now embarks on a similar quest with regard to the historiography and social science analyses of peasants. To this end, he attempts to breathe epistemological life into the plebeian/pastoral variant of the agrarian myth, and reifies the same essentialist images of the peasantry as those currently invoked by exponents of the 'new' populist postmodernism. Unlike the latter approach, however, which celebrates a peasantry mired in tradition and struggling to reproduce both it and their private property, Marxism locates peasants in the wider economic context, as part of a process that structures what they are, dividing them along class lines with different political interests to match.

This difference notwithstanding, Joyce dismisses the framework applied by Marxist theory to the study of peasants, banishing not just concepts such as class formation/consciousness/struggle, but also categories like modernity/development/progress. His search for an unmediated/pristine voice-from-below takes him instead to folklore as a viable source, despite its problematic methodology and theory. Contradictions abound, not least the smuggling by him into the narrative of the same tripartite classification applied by Lenin to the way capitalism differentiates the peasantry in terms of class position and

---

abstruse for all but the most patient and dedicated of readers. The gist of much of this literature is that peasanthood is not a changeless condition, and that peasants do not and always did not exist immobile in time'.

62   Joyce (2024: 338, n7).

interests, into rich/middle/poor strata. Ignored also is the current outcome of such a process: as part of a globally burgeoning industrial reserve, peasants wittingly or unwittingly assist employers to undermine worker solidarity, fuelling thereby the growth of reactionary populist ideology, and with it prolonging the life of the capitalist system.

CHAPTER 10

On the Continuing Necessity of (Marxist) Critique

ABUSES – see Tory Administration, *passim*.
GOVERNMENT – According to the Conservatives, the people should be defrauded, as much as possible, of the elective franchise, in order that an oligarchy may more easily defraud them of everything else.

> Two definitions of conservative government policy attributed to the humourist, poet, and novelist HORACE SMITH (1779–1849), written in 1836 but clearly no longer relevant today.[1]

∴

Introduction: Paradigms, Polemics, Popularity

The importance of Marxist critique to debates within the social sciences and development studies has been emphasized, rightly, in recent analyses.[2] That such observations are both timely and necessary is evident from the way in which a number of problematic developments now appear entrenched in academic discourse. First, strongly-made criticism itself – never mind that by Marxism – is now labelled 'polemical', and as such increasingly frowned upon. Second, concepts that are crucial to a Marxist framework, not to say Marxist theory itself, have been sidelined or banished from social science debates and development studies. And third, now what passes as Marxism has in many instances metamorphosed into 'Marxism'. It is these kinds of processes, their political effect, together with the reason for their emergence, that are considered briefly here.

Equally in need of emphasis are the two different forms of 'going back' arising from the unfolding of a critique. On the one hand, there is criticism based on rescuing/reinstating concepts and processes (for example, class differentiation, the industrial reserve army) that not merely belong to the framework in

---

1 Smith (1890: 7, 185).
2 Fasenfest (2022) and Das (2020).

question but are central to it, and that have been unjustifiably jettisoned in the name of 'newness', fashion or – much the same thing – the need to be seen to be 'popular'. On the other, by contrast, is the recuperation of concepts belonging to another framework altogether (for example, peasant essentialism, undifferentiated petty commodity production). This distinction, a crucial one epistemologically/politically, separates Marxist critique from various kinds of anti-Marxist equivalents: whereas the former sort of critique is engaged in restoring missing components from the *same* paradigm, the latter type of criticism has as its object the search for and installation of a wholly *different* paradigm.

It is a truism, therefore, that critique must be differentiated in terms of its place of origin on a left/right political spectrum. Whereas Marxism makes no secret of why, and on what basis, it enters into a debate, much non-/anti-Marxist theory does so in a variety of disguises, attempting thereby to hide and to 'normalise' its own epistemology. This process is compounded by an additional tendency, one that is as noticeable: subscribing uncritically to a paradigm widely (and inaccurately) labelled 'new' and 'radical', simply because it is novel and increasingly popular.[3] Gathering support, the paradigm snowballs: its claims remain unexamined, all the while assuming the status of a miracle cure guaranteed to confer immunity to Marxist 'infection', a form of exorcism periodically undertaken by bourgeois intellectuals. When the political effects of the latter shift are pointed out and criticized by Marxism, however, the frequent response in academia is offence, akin to *lèsé-majesté*.[4] This kind of reaction is perhaps no more than a symptom of the extent to which discussion in academia is now similar to that of a salon, governed by its codes and manners (polite conversation, eschewing controversy, avoiding politics).

This sort of response has to be linked in turn to the post-1960s fate of Marxist theory, and the way it vanished from the streets and into the academy. Over this period, many socialists had become not just ex-leftists but also senior academics, and to be reminded of how their earlier political allegiances had changed, and why, was clearly an uncomfortable experience. As revealing, therefore, are the reactions on the part of those criticized in this manner. Confronted with the extent to which they have strayed from Marxist theory and concepts, rather than addressing the critique the resort is frequently to nothing more than vituperation. The latter underlines the degree to which political debate in the academy has diminished, seen nowadays as 'unseemly' and quickly dismissed as 'polemical'. That such criticisms emanate from leftists is startling, not least

---

3 An instance of just this sort of process is considered elsewhere (Brass, 2023a).
4 On this kind of reaction to criticism, see Brass (2023b).

because historically the term 'polemical' – referring to fierce argumentation by the politically committed – was one used by opponents of Marx, Lenin, and Trotsky whenever any of the latter challenged bourgeois ideas.

A case in point is the novelist China Miéville, whose leftist contribution to a centenary discussion of the 1917 Russian Revolution was taken to task subsequently for being suffused by an undeniable aporia (a lukewarm 'defence' reproduced also in his latest book about the Communist Manifesto).[5] Hence a Marxist critique published later noted that, given the general hostility of newspaper articles marking the 1917 revolution, a more robust defence by Miéville would have helped.[6] This in turn elicited an irate response from Miéville himself, who rejected the critique as 'a strain of showboating machismo within the Left that treats consideration of any revolutionary parameters other than more or less precisely those of St Petersburg October 1917 ... as effete perfidy'.[7] It seems, therefore, that unease with revolutionary socialism is not confined to conservatives. On the nominal left, the idea of mass 'from below' agency designed to transcend capitalism by capturing its state also gives rise to misgiving.

The presentation which follows contains three sections, the first of which examines what connects academic opposition to Marxist theory/concepts, the emergence of what are claimed to be 'new' paradigms, and the role of fashion/popularity. Both the epistemology and the vehemence of anti-Marxist critiques are traced in the second, whilst a Marxist counter-critique is outlined in the third.

I

These days the sorts of judgement made by an editor of a major social science journal – as in the case, for example, of *Critical Sociology* – include having to form an opinion about the many and varied attempts to rewrite not just whole paradigms but also – and more insidiously – undermining longstanding epistemological and politically significant concepts on which these paradigms depend.

---

5   The contribution by Miéville to the 1917 centenary discussion is 'You say you want a revolution', *The Guardian* (London), 6 May 2017.
6   For the critique of Miéville, see Brass (2018a: 192). That the latter hit its target is evident from the response it elicited from Miéville (2022: 266–67 note 50), who complains about 'leftist revolutionism-policing' of Marxism, omitting to mention that the critique was aimed at him.
7   For this reaction, see Miéville (2022: 94), whose labelling of criticism as 'uncomradely' discloses two underlying assumptions. Not just the implied view that one should not question (a curious opinion for a leftist to hold), but also the extent to which academia ought to be no more than a present-day salon, engaged in polite conversation about unimportant subjects.

A Return to Yesterday

In terms of voice and platforms, the left appears superficially to be well served currently by publication outlets. On closer examination, however, this turns out to be deceptive. Although there are journals with words such as 'Left' and 'Marxism' in the title, legislative enforcement of an intellectual equivalent of Trades Description legislation would quickly require them to change their names. Thus, for example, the *New Left Review* has long since ceased to be a publication carrying serious discussion about any subject from a Marxist perspective, instead given these days to politically irrelevant salon discussions conducted by a coterie. Where the analysis of culture is concerned – and also, on occasion, politics – the *Financial Times* is more informative and more radical. The case of the journal *Rethinking Marxism* is, if anything, even worse, supporting Marxism in the same way as a noose supports a hanged man.[8]

Common to these journals is a misunderstanding as to the importance of the main concepts and systemic processes that are central to Marxist theory. One is provided with lots of favourable analysis on subjects like the body, sexuality, and non-class identities, together with the desirability of empowerment to be achieved within the confines of actually-existing neoliberalism, but little or nothing about class, its formation/consciousness/struggle under capitalism, and how the latter is connected to (or disconnected from) a socialist transition. The extent of the difficulty is evident from the way even basic concepts are being questioned or redefined so as no longer to be recognizable as such.

Where the history and presence of capitalism itself is concerned, this epistemological realignment has generated claims that can only be described as bizarre. Thus, for example, some now argue that as capitalism has in effect never existed, there is no such thing as capitalism in the sense Marx and Marxism understood.[9] By contrast, others maintain that capitalism has always existed throughout history, an ever-present and eternal systemic form.[10] To

---

[8] Other journals, that started out as Marxist or at least sympathetic to its political economy – among them *Economy and Society*, *International Review of Social History*, *The Journal of Peasant Studies*, The *Journal of Development Studies*, *Critique of Anthropology*, and *The Journal of Contemporary Asia* – have similarly abandoned this approach, to the extent of not publishing anything much that could be described as being informed by a Marxist framework. Yet others – such as *Science & Society* and *Capital & Class* – which are nominally still Marxist, do not appear any longer to endorse key concepts that are relevant to the core theory of its approach.

[9] See for example Jan Lucassen, whose views are examined in Chapter 5.

[10] This is the view of Jairus Banaji, on which see Brass (2021b: Chapters 5 and 6; 2022b: Chapter 3).

some degree, these two departures from Marxist theory coincide theoretically and politically. By extending what is termed capitalism in time and space, it ceases to be recognizable as the systemic form that Marx argued existed. To make it fit into these new times and spaces, therefore, requires a procrustean approach, discarding crucial elements and altering definitions, while at the same time adding 'new' components, concepts, and characteristics, invariably unconnected with – and in many cases opposed to – Marxism.

Furthermore, without some certainty about the contextually-specific existence of capitalism at a particular historical conjuncture, how is it any longer possible to talk about its transcendence, or – consequently – to formulate, advocate, and promote strategies designed to bring about a socialist transition? Crisis, exploitation, and oppression are as a result deemed 'natural' and thus unalterable aspects of the economy per se. The reason either for questioning the existence of capitalism, or for maintaining that it has always existed, appear to be the same. Accordingly, an effect of no capitalism, or alternatively its eternal character, is that there is consequently either no need to oppose it (since it doesn't exist) or its historical ubiquity forbids transcendence (since it cannot be eradicated). Either way, socialism is off the agenda, economically, politically, and ideologically.

New Paradigms, Old Assumptions

Declaring the 'old' redundant, supporters of the 'new' proclaim and celebrate its novelty amidst a growing popularity that confers on the approach a fashionable status among those conducting research.[11] Frequently, however, what is labelled the 'new' turns out to be nothing more than the resuscitation of an earlier social science or historical paradigm, itself the object of criticism by the 'old' approach now under challenge. Marxism makes no secret of its theoretical framework and epistemological lineage, originating in the nineteenth century. Unlike other paradigms, however, its claim to our attention lies not in being 'new', but rather in being politically relevant. This much is clear from what Marx had to say about the nature of popularity.

As pointed out by Liebknecht, Marx himself eschewed popularity, in the sense that he continued to pursue a radical political analysis in the face of criticism from bourgeois commentators ('phrase-mongers') who kept telling

---

11 How such claims to 'newness' are reproduced in academia is considered elsewhere (Brass, 2017b: Chapter 17; 2018a; 2018b).

him that such an approach was wrong.[12] In support of this persistence, in the Preface to the first edition of *Capital* Marx invoked the words of Dante, along the lines of 'Go your own way, and let others talk'.[13] The way Marx dismissed views advanced by 'adversaries of Socialism' designed to elicit 'the applause of the crowd', echoes later warnings made against uncritical endorsement of postmodernism, simply because it is academically fashionable.[14] Lenin and Trotsky were also chary about the courting of popularity, albeit in a different form, criticizing the espousal of leftist politics by *katheder*-socialists who then diluted Marxism in order to make it respectable in the eyes of the bourgeoisie. A similar view was expressed in 1960 by Ernesto Che Guevara, who referred to Cuban intellectuals supporting the Batista dictatorship as 'simple slavishness in the service of a disgraceful cause'.[15]

II

It is easy to overlook or ignore both the prevalence and the vehemence of anti-Marxist views – expressed not just by non-Marxists but also by some 'Marxists' – currently in circulation. Examined here, therefore, are three examples of this discourse, extending from a non-Marxist establishment historian (David Cannadine), via a couple of ex-Marxists (Ernesto Laclau, Chantal Mouffe), to a post-Marxist/non-Marxist social historian (Patrick Curry), who is influenced by Laclau. Whereas the former never was a Marxist, each of the latter four has seemingly metamorphosed from something akin to a fellow traveller into hostile critic. All the anti-Marxist views considered below demonstrate two things. First, the antagonistic relationship between Marxism and

---

12 See Liebknecht (1908: 84), who comments: 'Popularity being hateful to [Marx], he felt a holy wrath against soliciting popularity ... "Phrase-monger" was in his mouth the sharpest censure.' He continues (Liebknecht, 1908: 872): 'I have never forgotten the dangers of popularity; and if I remain unmoved ... by abusive language and the calumnies of our enemies – it is an art I learned from Marx'.

13 See Marx (1976: 93), who slightly altered the meaning of Dante's words (*'Vien retro a me, e lascia dir le genti'*).

14 'For popularity Marx entertained a sovereign contempt', notes Liebknecht (1908: 82), adding that '[a]nd while socialism has not spiritually soaked through the masses, the applause of the crowd can, as a logical consequence, be bestowed only on men belonging to no party or to the adversaries of Socialism'.

15 'The role intellectuals played here [in pre-revolutionary Cuba] was far less concealed than in Argentina', wrote Guevara (2022: 215) in a 1960 letter to the Argentinian novelist Ernesto Sábato, '[h]ere, the intellectuals were nothing but toadies, who did not disguise their real position as apathy like they did [in Argentina]. Moreover, they didn't even pretend to be intelligent. It was a matter of pure and simple slavishness in the service of a disgraceful cause, nothing more'.

academia, a long-standing and ongoing conflict determined by the political antimony of a bourgeois educational institution where Marxism is concerned. And second, although Marxists are frequently accused of harsh argumentation (= polemics) in defence of their views, and taken to task for the critical nature of their tone, less attention is paid to the abrasive tenor of criticism aimed at Marxism itself.

Class Dismissed

The antipathy expressed by Cannadine, in the form of concentrated vitriol he pours not just over Marxism generally but also over individual Marxists, is difficult to avoid.[16] Marxism is described by him variously as 'bravura, fortissimo exhortation', a 'crusading and coruscating polemic', while references to the proletariat in *The Communist Manifesto* are dismissed as 'a transparent falsehood' and 'naïve verbiage'.[17] Marx, Engels, and Lenin themselves fare no better, each of the former being described by Cannadine as 'privileged products' of the modern bourgeois world, and the latter is dismissed merely as a 'portentous ... self-styled revolutionary'.[18] The twofold inference is that being 'privileged products' eliminates any legitimacy for opposing the system, which is anyway fatuous when it occurs, amounting to no more than a pose (= 'self-styled revolutionary').[19]

Noting that others had allocated primacy to class earlier, but 'less stridently and polemically', Cannadine then asks whether 'class [has] ever been the most

---

16 Sources used by Cannadine for his critique of Marxism suggest that he tends to rely overmuch on publications and accounts by others who are either lukewarm about or similarly hostile to Marxism, giving rise to the issue of confirmation bias. These sources, both in the endnotes and the text itself, include not just Eduard Bernstein but also Walt Rostow, Hugh Trevor-Roper, Tristram Hunt, W.G. Runciman, Gareth Stedman-Jones, George Lichtheim, John H. Kautsky, Orlando Figes, and Richard Pipes. References to their criticisms are cited approvingly, as though they correspond to opinions held by neutral observers, whereas in fact they are for the most part ex-Marxists, anti-Marxists, and/or – like Cannadine himself – establishment historians.

17 Cannadine (2013: 94–95, 102, 107). Like others (see below) who announce the death of Marxist theory about class, Cannadine (2013: 129) too makes such a proclamation: 'As a preeminent form of human identity and the most significant category of historical explanation, class has had a great fall and, like Humpty Dumpty, it seems unlikely that the pieces will be put back together again anytime soon'.

18 Cannadine (2013: 100, 107, 110ff.).

19 The latter point, it should be noted, is the same kind of criticism as that made by Miéville (see above).

important and influential form of collective human identity and consciousness in the ways that Marx and Engels and their disciples, both practical and academic, repeatedly insisted that it was'.[20] This view of class as 'prime mover of change over time' could never work, he insists, because 'their attempts to explain ... all of human history on the basis of [classes] they believed to be in perpetual, sequential, and revolutionary conflict were deeply flawed'.[21] Denying the efficacy of class interests or consciousness, he invokes as an alternative the element of a 'common humanity', and laments that 'academic writing [supportive of this] has been produced by scholars whose interests are philosophical rather than historical'.[22] Given his position as part of the academic establishment, it is unsurprising that Cannadine holds the anti-Marxist views that he does; the same, however, cannot be said of others who target Marxism in a similar fashion.

Producing Curtains

As inescapable as that of Cannadine, therefore, is the vehemence with which Laclau dismisses Marxism. In a 1988 interview he began like the apostle Peter in the Garden of Gesthemane by observing that 'I have never been a "total" Marxist, someone who sought in Marxism a "homeland"; he continued by asserting that 'Marxism's destiny [may be no more than being] taken over by the boy scouts of small Trotskyist sects who will continue to repeat a totally obsolete language – and thus nobody will remember Marxism in twenty years' time'.[23] Marxism was not forgotten, and two decades later in a 2010 issue of the journal *Open* celebrating *The Populist Imagination*, Laclau and others endorse the importance of myth in the mobilization of support based on populism, as

---

20 See Cannadine (2013: 96), who – together with religion, nation, gender, race, and civilization – labels class as one of these 'divisive collective' identities that 'at worst [are] reductive and misleading', leading to 'polarizing modes of thought' (Cannadine, 2013: 7, 8, 9).

21 Cannadine (2013: 101).

22 See Cannadine (2013: 7, 102), who argues that 'changes in the economy were never so momentous, straightforward, or pervasive as to bring about those homogeneous, collective consciousness ... much less the perpetual conflict that Marx and Engels and their heirs said made history go'.

23 Laclau (1990: 178–79). In a similar vein, Mouffe (2018: 2) dismisses Marxism as 'class essentialism', on the grounds that its view that 'political identities were the expression of the position of social agents in the relations of production and their interests were defined by this position'. Wrongly, she concludes: 'It was no surprise that such a perspective [= Marxism] was unable to understand demands that were not based on "class"'.

indeed does the unforgotten Marxism. Whereas the latter sees this as negative, however, fomenting as it does false consciousness among workers, the former by contrast views it as positive. A succinct account of false consciousness is provided by Adorno, who refers to it as the effect of 'curtains' produced in order to obscure what it is possible to see beyond such drapes.[24]

Unlike Marxism, Laclau regards the discourse used by populism – nationalism, ethnicity – as a benign kind of 'from below' ideology, and thus progressive, not as evidence of 'from above' political manipulation.[25] Moreover, Laclau fails to situate the rise of populism within its economic context: namely, what are the economic reasons for the success of this ideology. Rather, he and others subscribe to a form of Sorelian instinctivism, or the 'feeling' that underlies and informs the reproduction of discourse based on myth. For its part, Marxism links the success of a populist mobilizing discourse privileging ideology informed by non-class identities (ethnicity, nationalism) to a globally expanding industrial reserve, its impact on labour market competition in metropolitan capitalism (in the form of immigration), together with the concern for their own livelihoods felt by those in employment or aspiring to this in receiving nations.

---

24 Writing in the immediate aftermath of the 1939–45 war, when the capacity of the Nazi regime to generate, reproduce, and consolidate an anti-semitic discourse required understanding, Adorno (1969: 661–62) contrasted 'political ignorance' with a generally high level of awareness 'in many other matters', observing: 'The ultimate reason for this ignorance might well be the opaqueness of the social, economic, and political situation to all those who are not in full command of all the resources of stored knowledge and theoretical thinking. In its present phase, our social system tends objectively and automatically to produce "curtains" which make it impossible for the naïve person really to see what it is all about. These objective conditions are enhanced by powerful economic and social forces which, purposely or automatically, keep the people ignorant. The very fact that our social system is on the defence, as it were, that capitalism, instead of expanding the old way and opening up innumerable opportunities to the people, has to maintain itself somewhat precariously and to block critical insights which were regarded as "progressive" one hundred years ago but are viewed as potentially dangerous today, makes for a one-sided presentation of the facts, for manipulated information, and for certain shift of emphasis which tend to check the universal enlightenment otherwise furthered by the technological development of communications'. He hastened to add that such political ignorance was not attributable to 'natural stupidity [or] a basic lack of the capacity for thinking'.

25 That populism is benign is a theme encountered in most of what he has published (e.g. Laclau, 2005a; 2005b). It is an influential view among those wishing to distance themselves from Marxism, particularly in the field of peasant studies (on which see Brass 2020, 2023a, 2023b).

Asked by an interviewer about what the latter terms 'ethno-populism', Laclau answers that '[e]thnic populism is important in Eastern Europe, but I don't think you will have a populism of that kind in Western Europe'.[26] From this Laclau misses the fact that discourse about ethnic 'otherness' is an ideology that operates at both a vertical and horizontal plane, and as such empowers and is empowered by populist discourse. In effect, Laclau makes the same error as that made earlier by Mudde, who asserted similarly in 2002 that populism 'plays a much more prominent role in contemporary Eastern European politics than in the West'.[27] The mistaken assumption in each case is that metropolitan capitalist nations are in some sense immune to political mobilization based on a discourse privileging ethnicity. Again, when asked by his interviewer how is it possible to describe neoliberalism as hegemonic, when in 1970s Latin America it was an economic project espoused by populist regimes but clearly imposed by 'those above' on the rest of society which opposed and resisted *laissez-faire*, Laclau responds weakly that neoliberalism 'was only hegemonic among economic and political elites', thereby undermining his own argument.[28]

Similarly problematic is the championing by Laclau and Mouffe of 'pluralistic' liberal social democratic regimes embodying what they term variously a 'national-populist tradition', a 'radicalization of democracy', or a 'radical and plural democracy', to be realized within and operated by parliamentary government.[29] In effect, theirs is an approach which leaves the capitalist system, together with its form of inequality, division of labour, class structure, income distribution, property relations, and its state intact.[30] This is because, as long

---

26 Laermans (2010: 74), who probes further by asking whether it is not the case that the anti-immigrant discourse of Dutch and French political right is also based on 'ethno-populism', to which Laclau gives an astonishingly bizarre reply. He rejects the premiss of the question, on the grounds that what is being said by the right is not that 'there exists a superior French [or Dutch] race', and that consequently the issue is 'not [about] ethno-populism but [rather] an anti-immigrant one'. Missed thereby is the fact that racism is not confined to ideas concerning superior/inferior position on an ethnic hierarchy, but also operates no less explicitly in cases where such identity is foregrounded without reference to hierarchical position.

27 See Mudde (2002: 231), whose erroneous interpretation of populism is analysed critically by Brass (2021b: 10–15).

28 Laermans (2010: 82). This does not prevent Mouffe (2018: 1) from asserting subsequently that 'we are witnessing a crisis of the neoliberal hegemonic formation', the inference being – once again – that neoliberal hegemony possesses a wider base of support than in fact it does.

29 These views are expressed by Laclau in his interview with Laermans (2010: 77) and by Mouffe (2018: 2).

30 That the existing state apparatus will remain in place is accepted by Mouffe (2018: 3).

as the rich and powerful remain unexpropriated, their capacity not just to frustrate or circumvent reform, and to roll back any attempts at economic change, but also to control politics and economic policy through a combination of state capture and/or evading state regulation, remains. It is perhaps a recognition of precisely this difficulty that lies behind the unwillingness of post-Marxism to indicate either what sort of agency populism entails, or indeed what sort of end would agency itself pursue.[31] Eschewing debate with critics, Mouffe adopts a *de haut en bas* attitude: 'I would like to make clear … that my aim is not to add another contribution to the already plethoric field of "populism studies" and I have no intention to enter the sterile academic debate about the "true nature" of populism'.[32]

Urgent Need of Renewal

Another symptomatic attack on Marxism is that by Curry, but now in the name of post-Marxist social history: it reveals the influence of Laclau and Mouffe, plus the epistemological route followed by social historiography, and why.[33] Defenders of Marxism are accused by Curry not only of 'undertaking a nostalgic retreat to the dogmas and elitism of teleology, reductionism and anachronism', but also of favouring 'empty and discredited nostrums'.[34] Dismissed by him is 'the assumption that the mode of production is a universal constant of history, and the working class therefore a privileged "universal" class', and Hobsbawm is praised for admitting in 1990 that 'the whole tradition dominated and inspired by the October Revolution has now come to an end'.[35]

---

31 Having proclaimed that '[t]he objective of a left populist strategy is the creation of a popular majority to come to power and establish a progressive hegemony', Mouffe (2018: 50) concedes that '[t]here is no blueprint for how this will take place or a final destination', adding that '[t]he same is true for the shape of the new hegemony that this strategy seeks to bring about'. In short, an admission that her post-Marxist populism has neither a concrete objective nor any idea how this might be achieved, were it to exist. Yet another example – were one needed – of postmodern aporia, embodied in utterances like 'struggles [will be] about different forms of subordination without attributing any *a priori* centrality to any of them'.

32 Mouffe (2018: 9).

33 The views expressed by Curry (1993) are regarded as symptomatic because they illustrate not just post-Marxist opposition to Marxism, but also – and more importantly – the theoretical underpinnings of this antagonism, and how these in turn licensed a slide into populism.

34 See Curry (1993: 172–73, 196 n58).

35 Curry (1993: 171). What is missed is that Hobsbawm was never really much of a Marxist (on which see Brass, 2021b: Chapter 2).

Much like Cannadine, Laclau, and Mouffe, therefore, Curry declares Marxism and class redundant, but with one difference: unlike each of the former, his post-Marxist version of social history emphasizes more strongly both the from-below empowerment conferred by non-class identity, and its political legitimacy (= an authentic grassroots voice).

Curry regards the anti-Marxist case made by J.C.D. Clark, a conservative anti-Marxist reinterpretation of seventeenth- and eighteenth-century British history, as anti-elitist, because it claimed not only that the *ancien régime* survived but also attributed this to popular/plebeian support for monarchy, aristocracy, and religion. This is because Clark, like E.P. Thompson, shared a rejection of economic determinism and placed an 'emphasis instead on taking people's beliefs seriously'.[36] Hence the conclusion by Curry that 'English Social History [based on a Marxist approach] is exhausted', while 'some of the recent criticism by the newer historians of the Right is apposite'.[37] Because of this combined misassessment (Marxist social historiography = wrong, conservative social historiography = correct), therefore, Curry then maintained that privileging a 'from below' voice in the manner effected by these rightwing historians consequently pointed 'in the direction of greater democratic pluralism in historical practice'. In short, a view – albeit a conservative one – that Curry regarded as not so different from that of E.P. Thompson, and like him opposed to what was termed Marxist 'economic reductionism' and 'classism'.[38] This 'democratic pluralism', Curry then argued, is the way forward politically, an historiographical approach 'in urgent need of renewal', by which he means getting rid of Marxist 'accumulation of reductionist residues' and replacing them with something different.

What form this alternative to Marxism is to take Curry makes only too clear in the following manner: 'My main purpose is to offer a prescription for renewal, centred on the post-Marxism of Laclau and Mouffe', extending further this political endorsement by adding that their 'work does indeed offer a potential renewal, no less in history writing than in other respects'.[39] The reason for this enthusiasm is, in his words, that 'hegemony offers a hopeful way

---

36 Curry (1993: 162).

37 Curry (1993: 178).

38 According to Curry (1993: 164–65), E.P. Thompson emphasized the 'cultural hegemony' exercised by rulers, maintaining that that plebeians were not unthinking accepters of their own subordination but bought into a world-view that reflected ruling class interests. His privileging of superstructural determination was criticized by other Marxists as 'culturalism'.

39 See Curry (1993: 172–73, 178), who defends both his attack on Marxism and his approval of Laclau and Mouffe by observing (original emphasis): '*This is precisely the critique of [Marxist] teleology, reductionism and anachronism – summed up in Laclau and Mouffe's*

forward [in the shape of] the recent post-Marxist interpretation of Ernesto Laclau and Chantal Mouffe'.[40] Not the least persuasive argument for Curry is that for Laclau a worker 'is no longer simply fundamentally that but embodies a number of other potentially equal important identities', the result being that 'the working class [is] no longer ... the necessary agent of global emancipation'.[41] In keeping with this view, Curry unsurprisingly defends postmodernism, denying that it is a form of relativism.[42] Rejecting the opposition expressed by some Marxists to what were now politically his new best friends, he goes on to lament 'the dreary abuse and denial that Laclau and Mouffe have elicited', evidence Curry thinks for 'the poverty of contemporary dogmatic Marxism'.[43]

What each of these critiques – by Cannadine, Laclau, Mouffe, and Curry – share is a discourse that urges historiography and/or social science generally to abandon Marxist theory once and for all. As such, it is an approach that invites – and should invite always – robust counter-critiques from those who remain Marxists.

III

Conversation, Collaboration, Cooperation?

Although right about declining academic interest in Marxism, in terms both of teaching and of studying, Cannadine fails to understand the cause. According to him, it was the result of disagreement among Marxists themselves about class, its presence and meaning, an intellectual fragility which in turn led to them being overwhelmed by non-Marxist historians.[44] Overlooked by him, however,

---

      work ... – that we found at the heart of Thompsonian social history, and restated (for different purposes) by Clark'.

40   Curry (1993: 167–68).

41   See Curry (1993: 170, 172), who notes that the decentring of class extended from Laclau and Mouffe to include also Stuart Hall, Gareth Stedman Jones, and Gavin Kitching.

42   Curry (1993: 176, 180).

43   Marxist critics named by Curry included Ellen Meiksins Wood, Norman Geras, and Bryan D. Palmer.

44   Acknowledging a renewed academic interest in Marxism over the 1960–1980 period, Cannadine (2013: 125–26) comments: 'While Marxist historians had captured the commanding heights of French academe, their British and American counterparts were less successful, especially in the universities of Oxford and Cambridge, or in the Ivy League, where class-based approaches remained essentially marginal, and where Marxism was never mainstream'. However, he misunderstands the reason for this: 'With Marxist historians unable to agree about the trajectory and trajectories of the classes they believed had existed in the past, it was scarcely surprising that scholars who did not share their faith

is the main reason for this waning interest: namely, that with the expansion of higher education from the 1960s onwards, Marxism (and Marxists) vanished from the street and into the universities, where it became a topic simply for study. The negative impact of this shift did not stop there, since it quickly became clear that to prosper in an anti-Marxist academic environment it was advisable quietly to jettison all things Marxist and socialist. Many leftists abandoned their political views (some turning into vehement anti-Marxists) and opted instead for the less threatening oppositional discourse then gaining hold, the 'new' populist postmodernism. This was the main reason why Marxism has declined: not because it was wrong, as Cannadine imagines, but rather because it some instances became an obstacle to becoming and remaining an academic.

Unknowingly, perhaps, his critique of Marxism is no different epistemologically from that made by Laclau and Mouffe, and for the same reason mistaken. 'For most people', argues Cannadine, 'work has only ever been part of their life … and has never been the sole determinant of how they see themselves or themselves in relation to others'.[45] Ignored thereby is that the main way in which people organize themselves and engage in agency is almost always connected to actual/potential threats to their livelihoods. That is, to protect or advance what are fundamentally economic interests (land, wages, employment), a process involving defence of or assault on property relations and institutions (the state) and structures (the law, media) connected to them. In short, action undertaken on the basis of class, promoting (or attacking) such interests as are central to this.[46]

To be sure, people do indeed have other identities and interests (for example, being competitors in pigeon racing contests, being members of a film society, supporting a particular football club) – characteristics Marxism has

---

were strongly critical of their overall approach, and since the 1980s the flow of criticism has swollen to a flood'.

45   Cannadine (2013: 103).

46   Downgrading the historical relevance of Marxist approaches featuring concepts like class and class struggle can only be made by ignoring their centrality to an understanding of much conflict taking place over the 20th century. Episodes such as the 1926 General Strike in the UK, via the divide and rule strategy pursued by the apartheid regime in South Africa (undermining previous black/white worker solidarity), the massacre of Communists in Shanghai during 1927 and Jakarta during 1965, to 'disappearances' of workers, trade unionists, and leftists generally by far-right military dictatorships in 1970s/1980s Chile, Uruguay, Argentina, and Brazil, are all instances of class struggle waged 'from above' (by the capitalist state representing the political interests of the rich and powerful) against 'those below' on account of the latter's *class* identity – that is, workers and/or poor peasants together with their organizations and political parties.

never denied – *but* the main kind of action undertaken (workers belonging to a trade union, joining demonstrations, or alternatively businessmen lobbying government) invariably affects what might be termed their *fundamental* interests: linked to the presence/absence of job/ livelihood prospects, the reproduction/dissolution of property rights, and thus to class position. In denying this, Cannadine sides with the post-Marxist interpretation of Laclau and Mouffe about 'equivalences', whereby a petition seeking a reduction in season ticket prices is for them no different from assembling on the street in order to oppose state oppression or a military coup, each in its own way being just another 'floating signifier' denoting a specific form of hegemony.[47]

Moreover, because he fails adequately to contextualize the decline of Marxism in academia, mistakenly attributing this to internal disagreement, Cannadine also misses the role and significance of wider societal determinants.[48] During the Thatcherite era, when *laissez-faire* was ideologically dominant, there was little institutional sympathy for leftist theory: funding provision both for university posts/courses and research projects exhibiting a specifically Marxist approach dropped noticeably. This, too, was a factor that contributed to declining interest in Marxism, leading to its replacement by the 'cultural turn'. Underestimating the acuteness of the class struggle 'waged from above' at this conjuncture, however, Cannadine egregiously declares Marx, Engels, and Lenin mistaken because 'capitalism has survived' on account of the fact that 'relations between the "proletariat" and the "bourgeoisie" have been characterized more ... by conversation, collaboration, and cooperation'.[49]

---

47   Thus, a landlord in Latin America or India might find himself in conflict with an alliance of tenants composed of rich peasants (small capitalists) and poor peasants (de facto workers), each of whom has a different agenda in class terms. Rich peasants want to acquire land ownership and freedom to sell its produce, while poor peasants want job security, higher wages, and better working conditions. In keeping with these different class agendas, once the landlord has been expropriated, and rich peasants become owners of the land they cultivate, as employers of labour-power they turn on their erstwhile allies whose demands are now aimed at them. In each stage of the struggle, therefore, agency by the landlord, the rich peasant, and the poor peasant is guided by their specific class identity and interests. In short, there is no escaping class, the wish by Cannadine to do so notwithstanding.

48   'The heroic narratives and broad generalizations that Marxist historians constructed have been overturned by the unprecedented research onslaught of the last twenty-five years', Cannadine (2013: 128) asserts confidently, 'which means it is no longer possible to view the past as a succession of gigantic Manichean encounters between rising, struggling, and falling classes'.

49   'Capitalism has survived, and with it the lumpenproletariat, the peasantry, and the petite bourgeoisie', states Cannadine (2013: 117), giving as the reason the following: 'While greed and exploitation persist, relations between the "proletariat" and the "bourgeoisie" have

Hegemonic Formation, Populist Moments, Floating Signifiers?

In the case of Laclau, what he cannot say is that there is indeed an additional dimension to populist discourse, one evident in the West, because to accept this would entail returning to the domain of Marxist theory and concepts, a prospect from which he is keen to distance himself. The dimension concerns the recognition that immigration is not merely an issue about national, ethnic, or cultural 'otherness' – as it is usually depicted – but also one of political economy. As Marx and Marxists have pointed out over the years, it is a question also of the industrial reserve, its enhancement, by whom or what, together with the reasons for this. As neoliberalism spreads over the globe, therefore, the inevitable result has been ever-more acute market competition, both between rival capitalists themselves and between the latter and their workers. It is an economic context that undermines consciousness and struggle based on class, and replaces them with twin forms of populism: the postmodern variant sees the cultural identity of the immigrant as empowering, on the basis of which open-door policy is to be supported politically, while the other, linked to varieties of 'nativist' conservativism, similarly privileges the nationality of the non-migrant worker.[50] Each rallies around the non-class identities wrongly perceived by post-Marxism as politically progressive.

The abandonment by Laclau and Mouffe of socialism as a desirable/feasible political objective, along with the proletariat as the universal historical subject, driving the process of class formation/consciousness/struggle, and Marxist theory in general, and its replacement with populist hegemony, leaves a space to be filled by an alternative subject and his/her agency. From this stemmed the post-Marxist claim that 'to counter the offensive of the right, it was crucial for Labour to expand its social basis ... and to incorporate the critics made by the new social movements, whose democratic demands it was essential to articulate'.[51] Into the space vacated by class Laclau and Mouffe placed a socio-economically undifferentiated category 'the people', an expansive catch-all identity beloved by populists, inside of which no significant contradiction was thought to exist.[52] In this post-Marxist framework every single occurrence – no

---

been characterized more in the long run by conversation, collaboration, and cooperation than by anger, antagonism, and animosity'. Painted is a scarcely recognizable history of capitalist development, a process seemingly devoid both of class and class struggle.

50   For details of these twin populisms, together with their deleterious impact on class mobilization and politics, see Brass (2022a: Chapter 7; 2022b: Chapter 1).
51   Mouffe (2018: 27, 28).
52   According to Mouffe (2018: 10–11, 24), for both her and Laclau populism is 'a discursive strategy of constructing a political frontier dividing society into two camps and calling

matter how trivial (a new way, perhaps, of combing one's hair) – becomes yet another instance of 'hegemonic formation' giving rise to its very own 'populist moment'.

Not the least of the many difficulties facing the post-Marxist populism of Laclau and Mouffe, therefore, is their interpretation of hegemony as a term lacking boundaries, exemplified by the concept 'floating signifier'. Having discarded Marxism and depriviliging class, they are unable to comprehend the way in which hegemony is structured – and thus limited – by class position, the latter bringing into consideration a positive/negative ideology that curbs any open-ended acceptance of hegemony. So, class, like Marxism itself, cannot be forgotten, in that the attempt to establish hegemony will always come up against hard economic facts. Thus, one cannot subscribe to any/all arguments/views regardless of political and economic interests (= the signifier on occasion refuses to float). Implausibly, however, the floating signifier of Laclau and Mouffe seems to indicate one can.

Accepting that the 1980s 'neoliberal hegemonic formation ... incorporated several themes of the counterculture', Mouffe fails to recognize its cause.[53] What the new social movements sought was autonomy, an ability to express individual choice as regards identity or lifestyle. This was an objective that was not just compatible with neoliberalism but actually central to capitalist development. In short, no contradiction existed between the neoliberal project of market expansion and the aims of the new social movements: not only were the latter neither radical nor a fundamental challenge to the system, but their desire to be accepted within it as presently constituted was much rather supportive of the accumulation process, providing it with an extra source of production so as to meet these additional kinds of consumer demand. Such compatibility notwithstanding, Mouffe appears puzzled by the fact that neoliberalism was capable of 'satisfying [new social movements] in a way that neutralizes their subversive potential', misunderstanding that – seen from the perspective of capital – these movements were never wholly 'subversive' in the first place.

---

for the mobilization of the "underdog" against "those in power"'. In a telling reveal, she then adds that '[i]t is not an ideology and cannot be attributed a specific programmatic content', but nevertheless operates contingently at specific conjunctures named by her as a 'populist moment'.

53 For this and what follows, see Mouffe (2018: 33–34).

Taking People's Beliefs Seriously?

Whilst Curry is right to argue that conservative social historiography set a trap for Marxism, ironically it is a trap into which he himself falls. This he does by signing up to all the familiar anti-Marxist tropes encountered in the conservative playbook over the years, in the belief that such backwards-looking ideology offers a politically viable alternative to Marxist theory. What is overlooked by Curry, therefore, is that 'taking people's beliefs seriously' amounts all too often to an uncritical endorsement of the 'from below' voice, simply because it is 'from below'. It is a methodology that licenses in turn the politically misleading *vox populi, vox dei* argument, on the grounds that 'there is nothing necessarily conservative or reactionary about those premises' because the latter are widely held grassroots beliefs.[54]

This approach is the cause of many of the difficulties faced by social history, equating as it has done – and does still – the 'from below' voice automatically with a progressive politics. Contrary to the view that interprets the 'from below' voice as an unmediated – and thus authentic – expression of grassroots political interests, more often than not it reproduces 'from above' ideology, resulting in the formation and consolidation of false consciousness. As such, many ideas/values which circulate at the grassroots – those which conservative and post-Marxist social historians wish to take seriously, as a 'from below' discourse – are much rather a form of class struggle waged 'from above'.[55]

Consequently, a methodologically uncritical approach to 'taking people's beliefs seriously', particularly where the element of class has been ruled out a priori, cannot but result in reification. However, swayed by a conservative historiography which methodologically based its political conclusions not on 'from above' ideology but rather on what 'those below' believed, Curry then proceeded to subscribe also to all the non-class identities that Marxism consigns to the realm of false consciousness.[56] Privileged analytically as a result of not addressing false consciousness, therefore, are not just the usual non-class forms of 'otherness' characterized by conservatives as innate, unchanging, and thus 'natural' and socio-economically non-transcendent – ethnicity, gender, patriotism, nationalism, and religion – but also 'the social history of marginalized "fringe" beliefs and people'.[57]

---

54 For this uncritical endorsement of the 'voice-from-below', see Curry (1993: 163–64).
55 Little wonder, therefore, that an establishment historian such as Cannadine (1997: 184ff.) expresses approval of social history, mistakenly labelling it 'socialist'.
56 Curry (1993: 180ff.).
57 Curry (1993: 179–80).

Significantly, Curry accepts that his enthusiasm for postmodernism and post-Marxism is tempered by its 'lending a perverse life to reaction'.[58] There is, however, nothing perverse about this connection: as has been outlined elsewhere, the link between on the one hand the 'new' populist postmodernism and post-Marxism, and on the other the political right, has deep historical roots in ideologies about Nature, the 'natural', and how these underwrite ideas about nation, tradition, race, hierarchy, and 'difference'.[59] All the latter are consistent with the arguments being deployed at that same conjuncture by the 'new' populist postmodernism, not just against Marxism as an organizing principle but also in opposition to modernity and Enlightenment discourse, grand narratives dismissed as inappropriate Eurocentric impositions on the rural 'other' in Third World nations.[60]

Conclusion

The decentring of Marxism that ends with a recuperation of populism is a trajectory that can be explained in a large part by reference to changes in the link between leftist politics and academic institutions; how Marxism vanished from the street into the academy. Not the least of the many problems this generated have been contradictions arising from having to change one's mind as a result of initially and uncritically espousing an in vogue 'new' paradigm becoming popular in terms of teaching/research, a result of initially not having asked the necessary questions about its politics and theory. Thus, a significant aspect of endorsing anti-Marxist/post-Marxist/populist discourses, together with the consequent discarding of Marxist concepts, is the nature and tone of the response to any criticism from inside the left that raises the issue of epistemological and political difficulties facing attempts to present such absences as compatible with Marxism.

---

58 Curry (1993: 180).
59 On this link, see Brass (2000).
60 Such views not only continue to inform his work but have in the process consolidated themselves. Accordingly, on his website – www.patrickcurry.co.uk – is found the following, dated 2019: 'They are all subjects that have been marginalised by, and within, mainstream modernity [which is] contemptuous of the wellsprings of life and its enchantment in the bodymind, the female, and the Earth'. About this approach he states: 'What I write out of, on the contrary, is "radical nostalgia" for what modernity mocks, marginalises, mimics and sometimes murders but which was good and worked, and (what is left of it) still is and still does. This, not reaction, is true conservatism of the kind espoused by Ruskin ... And in the empire of modernity, it is under assault'.

Defending Marxism against these kinds of academic critiques, both from outside (Cannadine) and from within (Laclau, Mouffe, Curry), frequently overlooks the chronology of such interventions. Usually, therefore, it is the latter sort of negative appraisal which initiates and sustains debate, in effect preceding any counter-arguments made by Marxists. Where/when the latter are missing, lukewarm, or deficient, however, Marxist theory is undermined, both epistemologically and politically. Moreover, since the claims informing the sort of critiques targeting Marxism are in many respects similar – questioning and/or rejecting not just class and struggle based on this but also modernity, development, and even capitalism itself, in favour of vague notions like populist 'hegemony' or 'common humanity' – they contribute to and at the same time consolidate both the influence and the fashionability/popularity exercised by anti-Marxist theory and politics. It is this deleterious process that underlines the continuing necessity of critique (or counter-critique) that is specifically Marxist.

Conclusion

Critiques presented here in defence of development range across a number of issues, all of which are central to discussions about the desirability or undesirability of this historical process. These include one particular aspect – labour market competition – of the debate about racism, why the reproduction of this ideology is more acute at some historical conjunctures but not others, the same question that can also be asked of the industrial reserve. Equally contentious is the current dominance of populist and postmodern interpretations of rural development, in the misleading guise of new paradigms, the object of which is to exorcise two ghosts: not just development itself, but also Marxist theory about this. The latter takes multiple forms, among them attempts to redefine capitalism as ever-present; simply to omit it conceptually from consideration; to rescue accumulation by separating it from the many accompanying political and economic contradictions; or, indeed, to incorporate in an unconvincing manner only those politically unthreatening elements of the original Marxist critique, resulting in a procrustean explanation of the development path.

Amidst widespread anxiety in academic circles that development is a luxury the world can no longer afford, the debate about this issue appears to be going backwards, towards one of two positions: either development as now constituted is all that society can ever hope for, or any development achieved hitherto must of necessity go into reverse. The impact of either position is profound, and its roots together with its effects are not difficult to discern. Contrary to the view that culture wars are waged *by* Marxism, therefore, it is evident that they are aimed much rather *at* Marxism, and through this against the notion of a socialist transition and the very idea of development itself. The consequences of the latter project, encompassing nineteenth century agrarian populism and its current reincarnation as postmodern theory, can be seen in the deleterious ideological role played by nationality/ethnicity in a variety of different historical contexts. These extend from the post-bellum American South, via the Brexit debate in the UK, to the privileging of identity politics in Western academia. Moreover, it is a trajectory in which rival interpretations – culture versus political economy – of labour market competition feature prominently.

Denial of development-as-modernity, a process leading ultimately to a socialist transition, which argues for the further systemic transformation of a progressive and non-exploitative kind, takes many forms by those who study this. Such denial can – and does – lead to an epistemological cul-de-sac, whereby the very possibility of systemic development is itself seen as impossible, since no one is certain any longer as to what the current mode – capitalism – is, let alone how it can be transcended, or, indeed, what social categories might have an interest in accomplishing this task. Advocating a more radical, Marxist interpretation of economic development, and no longer confining it simply to what occurred in Third World nations, has not been a popular view to hold, not least because it does not fit the prevailing narrative.

Central to these sorts of denial arguments is what might be termed the nothing-to-see-here view, a mainstream position in development studies which insists there is little or nothing problematic with the current system: one variant of this holds that what exists cannot be improved upon or transcended (because systemically it is historically eternal); another clings to the present, insisting that anyway there are no feasible alternatives, while yet others seek replacements that lie in the past, not the future. Hence the alarm of some in the 'development community' when not just unfree production relations, but also the industrial reserve, the agrarian myth, and populism itself – phenomena that in intellectual terms had been overlooked or downplayed where the study of capitalism was concerned – resurfaced in the midst of capitalism.

While class struggle was acknowledged to be an effect of capitalist development in Third World nations during the immediate post-war era, this was less so in the case of accumulation taking place in metropolitan contexts. In the latter case, those possessing means of production were perceived as unwilling – perhaps unable – to oppose policies like nationalisation, designed to bring industrial ownership under the control of the state, to be operated henceforth for the benefit of society as a whole. Missing from these sorts of analyses, which stressed not just the benign nature of capitalism and its state but also the ameliorating impact of social mobility and embourgeoisment, was a concept not just of class but of class *struggle*, and with it the rolling back of any progressive gains achieved hitherto. The lesson to emerge from that period was a simple one: unless the power of capital to strike back is eliminated – by, for example, expropriation without compensation of all its assets, the means of production/distribution/exchange – there would always be not just the possibility but also the probability that the rich and powerful would strike back, even if this required that they bide their time. This has been illustrated with reference to the industrial reserve.

In twenty-first century metropolitan capitalist nations, the issue of the industrial reserve takes on the form of immigration control, surfacing among the working class and its representatives as one of ethnic/national/gender identity. However, this ideological shift from class to non-class identity flourishes only so long as the continued reproduction of the system which gives rise to the industrial reserve army and its effects (low wages, fierce competition for jobs, and unemployment) is not addressed. In a capitalist context, therefore, socialists have on occasion argued that controlling the level of the industrial reserve army – opposing the continued access by employers to cheap migrant labour – diminishes competition in the labour market, thereby permitting workers and their representatives to begin to settle accounts with employers from a position of strength.

Where different ethnicities/nationalities compete for the same jobs, however, antagonism generated by such economic rivalry tends to be expressed in terms of these non-economic identities, a process that marginalizes (or eliminates) class unity and benefits employers. Marxism has always recognized this outcome and its attendant dangers, but unfortunately this is not true of a plethora of faux-Marxist or non-Marxist approaches that currently emanate from or adhere to a vague politics of human rights which simply advocate migrant empowerment regardless of anything else. Accepting the negative characterization by Marx of the industrial reserve, these contrasting approaches nevertheless in effect reverse this, and perceive its components not as undermining the gains made by labour but rather as the vanguard of the working class struggle against the capitalist system. Avoiding as they do any consideration of the economic implications of the industrial reserve for labour market competition, such notions play directly into the hands of capital and far-right populists.

Hence addressing the related issues of an unregulated expansion in the industrial reserve army and who benefits from this is a first step politically, after which – in a capitalist context – a government representing all workers (of whatever ethnicity and gender) can then proceed to implement regulation of wages and conditions. Accordingly, the twofold socialist object has to be: first, to protect the existing workforce, and – of course – migrants who are already part of this; and second, to oppose strongly any attempts by capital to add to their number. Whilst most, if not all, agree on the first object, there is much disagreement about the second. Organizing and supporting all components of the workforce is an obvious aim, and as such uncontroversial. However, agency of whatever kind directed towards this end cannot but be impeded by calls either for open borders or for border abolition.

The kinds of antagonism generated both in the post-bellum American South during the Reconstruction era and in the UK over Brexit illustrate the way an increase in labour market competition fuels the intensity and deleterious effect of racism, nationalism, and populism. This in turn underlines the gulf separating the economic approach on the part of Marxist theory to the industrial reserve army from that of postmodernism to the same process. Hence the negative impact on the political consciousness, solidarity, and organizational capacity among workers and poor peasants of competition for employment has long been recognized by Marxism, which argues that labour market competition empowers capital, enabling producers to operate the divide-and-rule tactic in order to pre-empt or undermine gains made by those of different ethnicities but united as workers.

This is a position that contrasts with the positive view regarding the desirability of open borders taken by postmodernism, a perception shared by capitalist producers. Seen mainly through cultural lens, the industrial reserve is interpreted by postmodern theory as evidence for progress, an assertion of the right to betterment on the part of the erstwhile colonized 'other'. In current discussion about development that invokes the trope of its being either fight or flight, postmodern theory aligns with the latter, celebrating this access to European and North American labour markets simply as a form of empowerment available to the Third World 'other'. Opposition to this process by those already in jobs, or aspiring to them, is decontextualized, and presented as evidence for nothing more than an innate racist ideology linked to Empire. Despite agreeing with much of the negative interpretation, liberals such as Beveridge – unlike Marxism – continued to see the market as playing a positive role in the development process.

A clue to the current absence from development studies of Marxist interpretations lies in the nature of academic life, and the kind of power and influence exercised via its institutional hierarchy, together with journal and book publishing linked to this. It was from within such contexts that the assault on leftist theory came, when after the 1960s Marxists were enrolled in university teaching posts. This element of hostility suffuses the narrative of the influential novel *The History Man*, the protagonist of which combined negative images of Marxism and sociology at the new universities in the person of a venal lecturer. The latter character is depicted as all-powerful, exercising hegemony over gullible and ill-informed students, notwithstanding the irony that Marxism was soon thereafter deprivileged in academic circles by the rise and fashionability in such contexts of the 'new' populist postmodernism.

The role of academic publication in ejecting not just Marxism but also the modernisation framework from development studies is evident in the trajectory followed by what is termed Critical Agrarian Studies, claimed by its exponents to provide a new and better interpretation of the dynamics and logics governing peasant economy and society. However, this model turns out to be the most recent rebirth of agrarian populism, a vehemently anti-Marxist theory that celebrates the inability of rural smallholders to transcend their non-class identity, its traditional customs and institutions, recasting both this primordial structure and its subject as innate, enduring, and empowering. As Marxists have pointed out on numerous occasions, by subsuming undifferentiated peasants and landless workers in the same category of 'the rural poor', such unquestioning adherence to populist ideology obscures the presence, operation, and effect of class divisions within the seemingly homogenous ranks of petty commodity production.

In another yet similar approach to the study of development, labelled by its exponents as 'global labour history', this process of disgorgement has metamorphosed yet further, capitalism as a systemic mode being declared by them either non-existent or historically ubiquitous. All the while, however, those making these arguments claim – with ever decreasing plausibility – that at heart they are still loyal Marxists, still using Marxist concepts, methods, and theory, and still arriving at socialist conclusions. The latter assertions notwithstanding, global labour historians urge the social sciences generally, and development studies in particular, to abandon Marxism, and instead adopt an all-inclusive notion of the extended working class mobilized against capital, incorporating – much like Critical Agrarian Studies – an undifferentiated peasantry, the lumpenproletariat, and petty traders. Overlooked thereby is that such components are as opposed to socialism as they are to capitalism.

Similarly devoid of socialism as a desirable/feasible outcome of the development process are other contributions to the debate examined here. Unable to recognize the roots of laissez-faire economics and populist ideology in what he terms classical liberalism, Fukuyama continues to promote unrealistically benign notions of viable political democracy as a justification for the continued relevance of the capitalist system. Equally problematic is the conceptualisation of unfree labour as neo-bondage, in terms of its sudden and mysterious origin, its confused epistemology, and its advocacy merely of a return to a caring/nice kind of accumulation. As worrying is the academic role of social history in advancing the cause of anti-Marxism: its practitioners have been responsible in a large part for the displacement of Marxism by the 'new' populist postmodernism, on the spurious grounds that recently discovered 'from below' discourse does not accord with class identity as argued by Marxism, ignoring thereby the efficacy of the way in which modern capitalism reproduces false consciousness.

Not the least important aspect of critique is the frequent disjuncture between the claims of theory about development and what its methodology permits. This can be illustrated with reference to the approaches of on the one hand both Joyce and van der Linden, each of whom privileges what fieldwork reveals without either of them having undertaken such research, and on the other Hart, whose own substantial and engaged fieldwork contributed to the formulation by him of the informal sector concept. The difficulties evident in the theory of both the former – based on attempts to deploy respectively a 'new' populist postmodern framework or global labour historiography to peasants and workers – derive from idealized notions undermined by evidence from anthropological research undertaken by others. This contrasts with Hart, from whose fieldwork emerged an important and enduring contribution to development theory. Ironically, whereas Joyce and van der Linden endorse a methodological approach they themselves lack, Hart is critical of the same kind of research procedure that he himself has undertaken.

Part of the difficulty is that the logic driving what are portrayed as 'new' paradigms in the social sciences is rarely problematized, let alone interrogated as to its political assumptions and agenda. It is clear, however, that one contributory factor is competition within academia for research funding: as competition intensifies, so the proliferation of 'new' interpretations or the discovery of 'new' areas for study proliferate. Brought into focus thereby is what are claimed overlooked issues which, so the argument goes, require further analysis before it is possible to be sure of anything linked to them. Social science discourse is currently awash with attempts to replace 'old' paradigms/concepts with 'new' equivalents, the latter gathering deserters from the former.

Moreover, since it is no longer possible to restrict the application of the term 'development' only to Third World countries on the global economic periphery, any approach nowadays must connect what happens there to what occurs in the core nations at the heart of the capitalist system. This has implications for the conceptual privileging of imperialism by some on the left who continue to promote the desirability not of socialism but rather more capitalism, on the spurious grounds that the latter still has a progressive role to play, and should be supported in doing so. Shifting the political focus in this manner, from advocating socialism and opposing capitalism to what amounts to its converse, resisting if not opposing socialism whilst favouring more accumulation, is perhaps the most egregious kind of difficulty facing those who continue to support the idea of development.

As a programme, it says a lot about the parlous state of leftist theory and practice. Given the nature of Marxist teleology, it ought to be impossible for socialists to disagree with the central argument that behind much anti-imperialism lies an unchallenged nationalist politics, which in the present age of transnational capital and class hides accumulation by and within non-imperialist countries; consequently, opposition to imperialism as oppressor of less powerful nations by more powerful ones tends to miss the full extent of capitalist development and class divisions operating across all globalized contexts. In one sense, it is difficult to believe that we are still having this debate, since its arguments, politics, and outcomes are issues that have a long history on the left.

Overlooking the politically crucial distinction between the anti-capitalism of the left (historical materialism) and the anti-capitalism of the right (populism) has in turn led to the misplaced advocacy by many on the left to espouse and mobilize simply on the basis of postmodern identity politics. That is to say, to endorse the agency of just about anyone, as long as it entails a form of cultural otherness (a different ethnicity, nationality, sexuality), in the process downgrading or discarding agency based specifically and only of those Marxism regards as belonging to the working class. Given the general absence of a radical socialist project – programme, policies, organization signalling the possibility of a systemic 'going beyond' what exists – one arrives at a political situation which currently is very familiar: national contexts that are marked by recurring economic crisis, increasing disenchantment with and opposition to capitalism, but in which any grassroots mobilization has nowhere to go but backwards, into the ranks of populism. The result, we know: having been deserted by the left, workers in large numbers have turned instead to populist solutions (privileging their national, not class identity) offered by the likes of Farage, Marine Le Pen, Trump, et al.

Unlike much current discussion about racism, characterized by adherents of the 'new' populist postmodernism as innate and linked solely and simply to nostalgia for empire, that about its role and reproduction in the American South of the late 1920s and early 1930s contextualizes racism in terms of acute labour market competition between blacks and whites in the economic downturn of the Depression era. This difference of emphasis is significant, in that it highlights the epistemology and politics involved: whereas the focus of the 'new' postmodern approach is on culture as the determinant of race and racism, an ideology that has no referent except to a history long past, that of analyses concerning a similar kind of 'otherness' in the South connects racism not only to slavery but also to the then-present exigencies of capital undergoing an economic crisis. It was the latter situation that ensured the ideological and political foregrounding of ethnic 'difference', a discourse that shifted the struggle away from the capital/labour relation of class and onto race, licensing thereby forms of control and oppression by one component of the working class over another in the search for jobs, a struggle that took the form of lynching.

Emerging from research conducted in 1930s American South is the finding that a crucial divide was not just the racist one between black and white (although that *was* important) but also within the ranks of the white population itself: between upper-class wealthy/landed whites and impoverished plebeian landless whites. This was reflected in their different attitudes towards blacks: whereas the landed/wealthy upper-class white viewed blacks through a benign, paternal lens, in effect casting them as subordinates who were innately inferior, lower-class whites who were landless extended their hostility to landless blacks and upper-class landed whites alike. The latter category was disliked for taking the sides of blacks against whites who were poor, in a sense underlining the extent to which landless blacks were perceived by white equivalents as being equal simply in terms of occupying the same *economic* position. As important in generating and reproducing these different attitudes towards blacks was the fact that wealthy whites did not have to compete with blacks in the same labour market, whereas poor whites did. Hence the antagonism on the part of the latter, both to their black rivals for jobs and tenancies, and to landed white employers for seeming to prefer these rival 'others' over those like themselves – whites who were poor.

As in the case of other, analogous contexts, what the racism of the 1930s American South confirms is that, by turning ethnically differentiated components of the working class against each other, it is possible to deflect the similarity of the impact on each component of the overarching economic crisis. In the 1930s, therefore, Southern planters restructured the agrarian labour process both in economic and in racial terms, whether of black by white or of white by black. That it is capitalism which drives racism, the crisis of the former enabling the reproduction of the latter kind of ideology, also underlines the negative outcome for class solidarity – and thus folly – of promoting or endorsing identity politics as the main (and perhaps only) form of struggle. Rather than being innate characteristics either of whites generally, or of white components of the working class, as argued by some exponents of the 'new' populist postmodernism, the idea and meaning of race can be – and frequently is – determined by the nature of the accumulation process.

Hence the outcome of legitimizing ethnic/national identity by postmodernism has been its impact on the capitalist labour market, simultaneously empowering new entrants while disempowering existing workforce components. Critique gives way to the celebration of 'otherness', overlooking thereby that racism as an ideology is combatted and undermined only when the economic conditions that give it life have themselves been eradicated, including crucially the process of labour market competition on which accumulation depends and thrives. While it is certainly true that not all instances of racist discourse can be attributed to this particular cause, it is equally true that explanations which do not address such rivalry miss an important factor contributing to the reproduction of this ideology. This lack is especially problematic when, as is now the case, the accumulation process together with its contradictions has spread across the world.

CONCLUSION 247

Among the hidden instances of labour market competition is film, where it is no longer possible for actors to play the part of a character whose ethnicity they do not share. By invoking cultural exclusivity, a group is able to limit to itself access to those eligible to portray characters like themselves on film. Thus, it would not be possible now for either Orson Welles or Laurence Olivier to play Othello, since to do so in terms of film – both the 1951 and 1965 version - they had to portray a different ethnicity: the essence of theatrical/cinematic acting is, after all, pretending to be someone else, portraying thereby a character that in most respects (historical, national, class, culture) is different from oneself. Although it is presented as culturally inappropriate to depict oneself as an 'other' that one is not ('you cannot portray me, since you are not of the same identity as I am'), there is a dimension to this repositioning that is rarely discussed: the opening up, if not the monopolization, of employment opportunities hitherto closed to particular groups (gender, race), limiting the theatrical/filmic representation of such characters to those like themselves.

The same could be said of representation in academia, where the prevailing ethos nowadays is that only the 'other' can talk about what 'otherness' means, a move away from the unconditional ability of all – regardless of identity – to participate in debate. This suggests that, along with Marxism and development, a wider sort of vanishing is now in play, a global phenomenon spreading throughout many aspects of capitalist society, replacing what were once thought to be desirable forms of advance – embodied, for instance, in the call for working class solidarity across nations, cultures, and ethnicities – with a backwards looking parochialism. In effect, closing down the broad idea of discussion aimed at realizing universal forms of identity and citizenship.

It is a process that makes difficult – if not shuts down – any moves towards the realization of internationalism, negating in particular the historical call for working class solidarity and organization across nations, cultures, and races. It is easy to spot – and agree about – biases informing, for example, nineteenth century accounts of colonization, structured as these are by Victorian notions regarding the immanence and immutability of racial hierarchy as a 'natural' phenomenon. Rather more difficult, and clearly open to disagreement, are the biases informing contemporary accounts of these same processes: namely, what are the politics of critiques aimed at colonialism/neo-colonialism and their attendant hierarchies that today emanate from erstwhile inhabitants of these same contexts. It could be argued that the latter discourse is to a large degree structured by nationalism, a politics that remains largely unquestioned and therefore un-interrogated. Whereas the ideological assumptions of the erstwhile colonizer is revealed for all to see and revile, those of the colonized – when made – are not.

Bibliography

Adorno, Theodor W., 1969 [1950] 'Politics and Economics in the Interview Material (Chapter XVII)', in Theodor W. Adorno, Else Frenkel-Brunswik, Daniel J. Levinson, and R. Nevitt Sanford (eds), *The Authoritarian Personality*, New York: W.W.Norton & Company Inc.

Akram-Lodhi, Haroon, and Cristóbal Kay, 2021, 'The Diversity of Classical Agrarian Marxism', in Haroon Akram-Lodhi *et al.* (eds), *Handbook of Critical Agrarian Studies*.

Akram-Lodhi, Haroon, Kristina Dietz, Bettina Engels, and Ben M. McKay (eds.), 2021a, *Handbook of Critical Agrarian Studies*, Cheltenham: Edward Elgar Publishing.

Akram-Lodhi, Haroon, Kristina Dietz, Bettina Engels, and Ben M. McKay 2021b, 'An Introduction to the *Handbook of Critical Agrarian Studies*', in Haroon Akram-Lodhi *et al.* (eds) *Handbook of Critical Agrarian Studies*.

Amin, Shahid, and Gautam Bhadra, 1994, 'Ranajit Guha: A Biographical Sketch', in David Arnold and David Hardiman (eds.), *Subaltern Studies VIII: Essays in Honour of Ranajit Guha*, Delhi: Oxford University Press.

Amis, Kingsley, 1970, *What Became of Jane Austin and Other Questions*, London: Jonathan Cape.

Angelo, Larian, 1997, 'Old Ways in the New South: The Implications of the Recreation of an Unfree Labor Force', in Tom Brass and Marcel van der Linden (eds.), *Free and Unfree Labor: The Debate Continues*, Berne: Peter Lang AG.

Arnold, David, 1984, 'Gramsci and Peasant Subalternity in India', *The Journal of Peasant Studies*, Vol. 11, No. 4.

Augstein, Hannah Franziska, 1996, *Race: The Origins of an Idea, 1760–1850*, Bristol: Thoemmes Press.

Ayers, Edward L., 1984, *Vengeance and Justice: Crime and Punishment in the 19th Century American South*, New York: Oxford University Press.

Balibar, Étienne, Sandro Mezzadra, and Ranabir Samaddar (eds.), 2012, *The Borders of Justice*. Philadelphia: Temple University Press.

Banaji, Jairus, 2003, 'The Fictions of Free Labour, Contract, Coercion, and So-called Unfree Labour', *Historical Materialism*, Vol. 11, No. 3.

Banaji, Jairus, 2010, *Theory as History: Essays on Modes of Production and Exploitation*, Leiden: Brill.

Barker, Ernest, 1951, *Principles of Social and Political Theory*, New York: Oxford University Press.

Barran, D.H., Harry G. Johnson, George Rowland, and the Earl of Cromer, 1969, *Rebuilding the Liberal Order*, London: Institute of Economic Affairs.

Basso, Pietro, 2021, 'Marx on Migration and the Industrial Reserve Army: Not to be Misused!', in Marcello Musto (ed.), *Rethinking Alternatives with Marx*, London: Palgrave Macmillan.

Bauer, Arnold J., 1979, 'Rural Workers in Spanish America: Problems of Peonage and Oppression', *Hispanic American Historical Review*, Vol. 59, No. 1.

Bayly, C.A., 1988, 'Rallying Round the Subaltern', *The Journal of Peasant Studies*, Vol. 16, No. 1.

Beck, E.M., and Stewart E. Tolnay, 1992, 'A Season for Violence: The Lynching of Blacks and Labor Demand in the Agricultural Production Cycle in the American South', *International Review of Social History*, Vol. 37, No. 1.

Beck, Ulrich, and Johannes Willms, 2004, *Conversations with Ulrich Beck*, Cambridge: Polity Press.

Benn, Tony, 2007, *More Time for Politics: Diaries 2001–2007* (Selected and Edited by Ruth Winstone), London: Hutchinson.

Bernstein, Henry, 1990, 'Taking the Part of Peasants', in Henry Bernstein, Ben Crow, Maureen Mackintosh, and Charlotte Martin (eds.), *The Food Question*, London: Earthscan.

Bernstein, Henry, 1996/97, 'Agrarian Questions Then and Now', *The Journal of Peasant Studies*, Vol. 24, Nos. 1–2.

Bernstein, Henry, 2006, *Is there an Agrarian Question in the 21st Century?* Paper presented at CASID Congress, Toronto, June 1–3.

Bernstein, Henry, 2010, *Class Dynamics of Agrarian Change*, Halifax and Winnipeg: Fernwood Publishing.

Bernstein, Henry, 2013, 'Agriculture, Class and Capitalism', *International Socialism*, No. 138.

Bernstein, Henry, 2018, 'The "Peasant Problem" in the Russian Revolution(s), 1905–1929', *The Journal of Peasant Studies*, Vol. 45, Nos. 5–6.

Bernstein, Henry, 2021a, 'Russian to Modern World history: Teodor Shanin and Peasant Studies', *Journal of Agrarian Change*, Vol. 21, No. 1.

Bernstein, Henry, 2021b, 'Into the Field with Marx: Some Observations on Researching Class', in Alessandra Mezzadri (ed.), *Marx in the Field*, London: Anthem Press.

Bernstein, Henry, 2022, 'Review of Handbook on Urban Food Security in the Global South', *Journal of Agrarian Change*, Vol. 22, No. 4.

Bernstein, Henry, 2023, 'JPS at 50: Some Personal Reminiscences', *The Journal of Peasant Studies*, Vol. 50, No. 2.

Bernstein, Henry, and T.J. Byres, 2008, 'Foreword', in Saturnino M. Borras, Marc Edelman, and Cristóbal Kay (eds.), *Transnational Agrarian Movements Confronting Globalization*, Chichester: Wiley.

Bernstein, Henry, Harriet Friedmann, Jan Douwe van der Ploeg, Teodor Shanin, and Ben White, 2018, 'Forum: Fifty Years of Debate on Peasantries, 1966–2016', *The Journal of Peasant Studies*, Vol. 45, No. 4.

Berry, Sara, 2021, 'Class', in Haroon Akram-Lodhi *et al.* (eds) *Handbook of Critical Agrarian Studies*.

Beveridge, William H., 1931a, *Unemployment: A Problem of Industry (1909 and 1930)*, London: Longmans, Green and Co.

Beveridge, William H., 1931b, *Tariffs: The Case Examined*, London: Longmans, Green and Co.

Beveridge, William H., 1943, *Pillars of Security, and Other War-time Essays and Addresses*, London: George Allen & Unwin Ltd., Museum Street.

Beveridge, William H., 1944, *Full Employment in a Free Society*, London: George Allen & Unwin Ltd.

Beveridge, William H., 1946, *Why I am a Liberal*, London: Herbert Jenkins Limited.

Beverley, John, 2004, 'Subaltern Resistance in Latin America: A Reply to Tom Brass', *The Journal of Peasant Studies*, Vol. 31, No. 2.

Bhopal, Kalwant, 2018, *White Privilege: The Myth of a Post-Racial Society* (Foreword by Yasmin Alibhai-Brown), Bristol: Policy Press.

Blok, Anton, 1988 [1974], *The Mafia of a Sicilian Village 1860–1960: A Study of Violent Peasant Entrepreneurs*, Cambridge: Polity.

Bodley, John H., 1982, *Victims of Progress*, Palo Alto, CA: Mayfield Publishing Company.

Borges, Jorge Luis, 1998, 'The Ethnographer', *Collected Fictions* (Translated by Andrew Hurley), New York: Penguin Putnam, Inc.

Borras, Saturnino, 2023, 'Politically Engaged, Pluralist and Internationalist: Critical Agrarian Studies Today', *The Journal of Peasant Studies*, Vol. 50, No. 2.

Bradbury, Malcom, 1962, *All Dressed Up and Nowhere to Go: The Poor Man's Guide to the Affluent Society*, London: Max Parrish & Co. Ltd.

Bradbury, Malcom, 1975, *The History Man*, London: Secker & Warburg.

Bradbury, Malcom, 1983, *Rates of Exchange*, London: Secker & Warburg.

Bradbury, Malcom, 1986, *Why Come to Slaka?* London: Secker & Warburg.

Bradbury, Malcom, 1987, *No, Not Bloomsbury*, London: André Deutsch Limited.

Bradbury, Malcom, 2006, *Liar's Landscape: Collected Writing from a Storyteller's Life* (Edited by Dominic Bradbury, with an afterword by David Lodge), London: Picador.

Bradley, Gracie Mae, and Luke de Noronha, 2022, *Against Borders: The Case for Abolition*, London: Verso.

Brass, Tom, 1980, 'Class Formation and Class Struggle in La Convención, Peru', *The Journal of Peasant Studies*, Vol. 7, No. 4.

Brass, Tom, 1983a, 'Agrarian Reform and the Struggle for Labour-Power: A Peruvian Case-Study', *The Journal of Development Studies*, Vol. 19, No. 3.

Brass, Tom, 1983b, 'Of Human Bondage: Campesinos. Coffee and Capitalism on the Peruvian Frontier', *The Journal of Peasant Studies*, Vol. 11, No. 1.

Brass, Tom, 1986a, 'Debt Bondage in India', in Raana Gauhar (ed.), *Third World Affairs 1986*, London: Third World Foundation for Social and Economic Studies.

Brass, Tom, 1986b, 'Unfree Labour and Capitalist Restructuring in the Agrarian Sector: Peru and India', *The Journal of Peasant Studies*, Vol. 14, No. 1.

Brass, Tom, 1988, 'Slavery Now: Unfree Labour and Modern Capitalism', *Slavery & Abolition*, Vol. 9, No. 2.

Brass, Tom, 1990, 'Class Struggle and the Deproletarianisation of Agricultural Labour in Haryana (India)', *The Journal of Peasant Studies*, Vol. 18, No. 1.

Brass, Tom, 1991, 'Moral Economists, Subalterns, New Social Movements, and the (Re) Emergence of a (Post) Modernised (Middle) Peasant', *The Journal of Peasant Studies*, Vol. 18, No. 2.

Brass, Tom, 1995, 'Reply to Utsa Patnaik: If the Cap Fits', *International Review of Social History*, Vol. 40, No. 1.

Brass, Tom, 1997a, 'The Agrarian Myth, the 'New' Populism and the 'New' Right', *The Journal of Peasant Studies*, Vol. 24, No. 4.

Brass, Tom, 1997b, 'Immobilized Workers, Footloose Theory', *The Journal of Peasant Studies*, Vol. 24, No. 4.

Brass, Tom, 1999, *Towards a Comparative Political Economy of Unfree Labour: Case Studies and Debates*, London: Frank Cass Publishers.

Brass, Tom, 2000, *Peasants, Populism, and Postmodernism: The Return of the Agrarian Myth*, London: Frank Cass Publishers.

Brass, Tom, 2000a, 'Unmasking the Subaltern, or Salamis without Themistocles', *The Journal of Peasant Studies*, Vol. 28, No. 1.

Brass, Tom, 2002, 'On Which Side of What Barricade? Subaltern Resistance in Latin America and Elsewhere', *The Journal of Peasant Studies*, Vol. 29, Nos. 3–4.

Brass, Tom, 2003, 'Why Unfree Labour is Not "So-Called": The Fictions of Jairus Banaji', *The Journal of Peasant Studies*, Vol. 31, No. 1.

Brass, Tom, 2004, '"Medieval Working Practices"? British Agriculture and the Return of the Gangmaster', *The Journal of Peasant Studies*, Vol. 31, No. 2.

Brass, Tom, 2005, 'The Journal of Peasant Studies – The Third Decade (1993–2004)', *The Journal of Peasant Studies*, Vol. 32, No. 1.

Brass, Tom, 2007, 'Weapons of the Week, Weakness of the Weapons: Shifts and Stasis in Development Theory', *The Journal of Peasant Studies*, Vol. 34, No. 1.

Brass, Tom, 2011, *Labour Regime Change in the Twenty-First Century: Unfreedom, Capitalism and Primitive Accumulation*, Leiden: Brill.

Brass, Tom, 2013, 'Good Companions or Usual Suspects?', *Capital & Class*, Vol. 37, No.1.

Brass, Tom, 2014a, 'Debating Capitalist Dynamics and Unfree Labour: A Missing Link?', *The Journal of Development Studies*, Vol. 50, No. 4.

Brass, Tom, 2014b, *Class, Culture, and the Agrarian Myth*, Leiden: Brill.

Brass, Tom, 2015a, 'Peasants, Academics, Populists: Forward to the Past?', *Critique of Anthropology*, Vol. 35, No. 2.

Brass, Tom, 2015b, 'Free Markets, Unfree Labour: Old Questions Answered, New Answers Questioned', *Journal of Contemporary Asia*, Vol. 45, No. 3.

Brass, Tom, 2017a, 'Who These Days is Not a Subaltern? The Populist Drift of Global Labour History', *Science & Society*, Vol. 81, No. 1.

Brass, Tom, 2017b, *Labour Markets, Identities, Controversies: Reviews and Essays, 1982–2016*. Leiden: Brill.

Brass, Tom, 2018a, *Revolution and Its Alternatives: Other Marxisms, Other Empowerments, Other Priorities*. Leiden: Brill.

Brass, Tom, 2018b, 'The Incompatibles? Socialism, Academia, Populism', *Populism*, Vol. 1, No. 2.

Brass, Tom, 2020, 'Is Agrarian Populism Progressive? Russia Then, India Now', *Critical Sociology*, Vol. 46, Nos. 7–8.

Brass, Tom, 2021a, 'Neo-populist Fables: The Other World of A.V. Chayanov', *Critical Sociology*, Vol. 48, Nos. 7–8.

Brass, Tom, 2021b, *Marxism Missing, Missing Marxism: From Marxism to Identity Politics and Beyond*. Leiden: Brill.

Brass, Tom, 2022a, 'Great Replacement and/as the Industrial Reserve: Populism or Marxism?', in David Fasenfest (ed.), *Marx Matters*, Leiden: Brill.

Brass, Tom, 2022b, *Transitions: Methods, Theory, Politics*. Leiden: Brill.

Brass, Tom, 2022c, 'Twisted Trajectories, Curious Chronologies: Revisiting the Unfree Labour Debate', *Critical Sociology*, Vol. 48, No. 1.

Brass, Tom, 2022d, 'Marxism, Peasants, and the Cultural Turn: The Myth of a "Nice" Populism', in David Fasenfest (ed.), *Marx Matters*, Leiden: Brill.

Brass, Tom, 2023a, 'Critical Agrarian Studies as Populist Land-Grab', *Critical Sociology*, Vol. 49, No. 3.

Brass, Tom, 2023b, 'Academia, Marxism, and Sociology: A Warning from "The History Man"', *Class, Race and Corporate Power* 11(2): Article 10. Available at: https://digitalcommons.fiu.edu/classracecorporatepower/vol11/iss2/10.

Brass, Tom, 2023c, 'Introspective Anthropology? Talking About Me (and You)', *Critical Sociology*, Vol. 49, Nos. 7–8.

Brass, Tom, 2023d, 'Defending Liberalism, Promoting Capitalism: Fukuyama's Scylla and Charybdis', *Critical Sociology*, Vol. 49, No. 2.

Brass, Tom, 2023e, 'Communication: Marxism, Deproletarianisation, Geographers and Non-geographers', *Human Geography*, Vol. 16, No. 1.

Brass, Tom, and Henry Bernstein, 1992, 'Introduction: Proletarianisation and Deproletarianisation on the Colonial Plantation', *The Journal of Peasant Studies*, Vol. 19, Nos. 3–4.

Brass, Tom, and Marcel van der Linden (eds.), 1997, *Free and Unfree Labour: The Debate Continues*, Bern: Peter Lang.

Breman, Jan, 1974, *Patronage and Exploitation: Changing Agrarian Relations in South Gujarat, India*, Berkeley, CA: University of California Press.

Breman, Jan, 1978–79, 'Seasonal Migration and Co-operative Capitalism: The Crushing of Cane and of Labour by the Sugar Factories of Bardoli, South Gujarat, Parts 1 and 2', *The Journal of Peasant Studies*, Vol. 6, No. 1 and Vol. 6, No. 2.

Breman, Jan, 1985, *Of Peasants, Migrants and Paupers*, Delhi: Oxford University Press.

Breman, Jan, 1989, 'The Disintegration of the *Hali* System', in Hamza Alavi and John Harriss (eds.), *Sociology of 'Developing Societies': South Asia*, London: Macmillan.

Breman, Jan, 1990, '"Even Dogs are Better Off": The Ongoing Battle Between Capital and Labour in the Cane-fields of Gujarat', *The Journal of Peasant Studies*, Vol. 17, No. 4.

Breman, Jan, 1993, *Beyond Patronage and Exploitation: Changing Agrarian Relations in South Gujarat*, Delhi: Oxford University Press (a new edition of Breman, 1974).

Breman, Jan, 1996, *Footloose Labour: Working in India's Informal Economy*, Cambridge: Cambridge University Press.

Breman, Jan, 1999a, 'Industrial Labour in Post-Colonial India – I: Industrialising the Economy and Formalizing Labour', *International Review of Social History*, Vol. 44, No. 3.

Breman, Jan, 1999b, 'Industrial Labour in Post-Colonial India – II: Employment in the Informal-Sector Economy', *International Review of Social History*, Vol. 44, No. 3.

Breman, Jan, 2007, *Labour Bondage in West India: From Past to Present*, New Delhi: Oxford University Press.

Breman, Jan, 2013, *At Work in the Informal Economy of India*, New Delhi: Oxford University Press.

Breman, Jan, 2016, *On Pauperism in Present and Past*, New Delhi: Oxford University Press.

Breman, Jan, 2019, *Capitalism, Inequality and Labour in India*, Cambridge: Cambridge University Press.

Breman, Jan, 2023, *Fighting Free to Become Unfree Again: The Social History of Bondage and Neo-Bondage of Labour in India*, New Delhi: Tulika Books.

Breman, Jan, 2024, *Colonialism, Capitalism and Racism*, Amsterdam: Amsterdam University Press.

Bruno, Sergio, 1979, 'The Industrial Reserve Army, Segmentation and the Italian Labour Market', *Cambridge Journal of Economics*, Vol. 3, No. 2.

Burgess, Anthony, 1984, *Ninety-Nine Novels: The Best in English since 1939, a Personal Choice by Anthony Burgess*, London: Alison & Busby Limited.

Byres, T.J., 1996, *Capitalism from Above and Capitalism from Below*, London: Macmillan.

Cannadine, David, 1997, *The Pleasures of the Past*, Harmondsworth: Penguin Books.

Cannadine, David, 2013, *The Undivided Past: History Beyond our Differences*, London: Allen Lane, Penguin Books.

Caponigri, A.R., 1955, *History and Liberty: The Historical Writings of Benedetto Croce*, London: Routledge and Kegan Paul.

Capps, Gavin, and Liam Campling, 2016, 'An Interview with Henry Bernstein', *Journal of Agrarian Change*, Vol. 16, No. 3.

Chaturvedi, Vinayak (ed.), 2012, *Mapping Subaltern Studies and the Postcolonial*, London: Verso.

Chayanov, A.V., 1966 [1923], *The Theory of Peasant Economy* (Edited by Daniel Thorner, Basile Kerblay, and R.E.F. Smith), Homewood, Illinois: Published for The American Economic Association by Richard D. Irwin, Inc.

Chayanov, A.V., 1991 [1926] *The Theory of Peasant Cooperatives*, London and New York: I.B. Tauris & Co. Ltd.

Chesterton, Cecil, 1940 [1919] *A History of the United States*, London: J.M. Dent & Sons, Ltd.

Christian, Cornelius, 2017, 'Lynchings, Labour, and Cotton in the US South: A Reappraisal of Tolnay and Beck', *Explorations in Economic History*, Vol. 66.

Clark, Alan, 2010, *Alan Clark – A Life in His Own Words: Diaries 1972–1999* (Edited by Ion Trewin), London: Orion Publishing Co.

Cockburn, Alexander, and Robin Blackburn, 1969, *Student Power: Problems, Diagnoses, Action*. Harmondsworth: Penguin Books.

Cohn-Bendit, Daniel, and Gabriel Cohn-Bendit, 1968, *Obsolete Communism: The Left-wing Alternative*, New York: McGraw-Hill Book Company.

Collins, Jock, 1984, 'Marx's Reserve Army: Still Relevant 100 Years On', *The Journal of Australian Political Economy*, Vol. 16 (March).

Cox, C.B., and A.E. Dyson (eds.), 1969–77, *Black Papers on Education*, London: The Critical Quarterly.

Croce, Benedetto, 1941, *History as the Story of Liberty* (Translated by Sylvia Sprigge), London: George Allen and Unwin Limited.

Croll, Andy, 2002, 'The Impact of Postmodernism on Modern British Social History'. *Moving the Social*, Vol. 27.

Crossman, R.H.S., 1965, 'The Lessons of 1945', in Perry Anderson and Robin Blackburn (eds.), *Towards Socialism*, London: The Fontana Library.

Crossman, R.H.S., 1979, *The Crossman Diaries: Selection from 'The Diaries of a Cabinet Minister' 1964–1970* (Edited by Anthony Howard), London: Magnum Books.

Curry, Patrick, 1993, 'Towards a Post-Marxist Social History: Thompson, Clark and beyond', in Adrian Wilson (ed.), *Rethinking Social History: English Society 1570–1920 and Its Interpretation*, Manchester: Manchester University Press.

Curtis, Bruce, 2015, 'From an Anti-Podium', *Sociology*, Vol. 49, No. 6.

Daniel, Pete, 1972, *The Shadow of Slavery: Peonage in the South 1901–1969*, Urbana, IL: University of Illinois Press.

Das, Raju J., 2014, *A Contribution to the Critique of Contemporary Capitalism: Theoretical and International Perspectives*, New York: Nova Science Publishers.

Das, Raju J., 2017, *Marxist Class Theory for a Skeptical World*, Leiden: Brill.

Das, Raju J., 2020, 'On the Urgent Need to Re-Engage Classical Marxism', *Critical Sociology*, Vol. 46, Nos. 7–8.

de Janvry, Alain, and Elisabeth Sadoulet, 2011, 'Subsistence Farming as a Safety Net for Food-price Shocks', *Development in Practice*, Vol. 21, Nos. 4–5.

Deas, Malcom, 1977, 'A Colombian Coffee Estate: Santa Bárbara, Cundinamarca, 1870–1912', in Kenneth Duncan, Ian Rutledge, and Colin Harding (eds.), *Land and Labour in Latin America*, Cambridge: Cambridge University Press.

Desai, Vandana, and Robert B. Potter (eds.), 2008, *The Companion to Development Studies*, London: Hodder Education.

Dietz, Kristina, 2021, 'Political Ecology', in Haroon Akram-Lodhi *et al.* (eds) *Handbook of Critical Agrarian Studies*.

Dixon, Thomas Jr., 1889, *Living Problems in Religion and Social Science*, New York: Charles T. Dillingham, Broadway.

Dixon, Thomas Jr., 1903a, *The Leopard's Spots*, New York: Doubleday, Page & Co.

Dixon, Thomas Jr., 1903b, *The One Woman: A Story of Modern Utopia*, New York: Doubleday, Page & Co.

Dixon, Thomas Jr., 1905, *The Life Worth Living*, New York: Doubleday, Page & Co.

Dixon, Thomas Jr., 1909, *Comrades: A Story of Social Adventure in California*, New York: Doubleday, Page & Co.

Dixon, Thomas Jr., 1911, *The Root of Evil*, New York: Doubleday, Page & Co.

Dixon, Thomas Jr., 1941 [1905], *The Clansman*, New York: Triangle Books.

Dobb, Maurice, 1955, 'Full Employment and Capitalism [1950]', in *On Economic Theory and Socialism: Collected Papers*, London: Routledge & Kegan Paul.

Drage, Geoffrey, 1904, *Russian Affairs*, London: John Murray.

Draper, Theodore, 1957, *The Roots of American Communism*, New York: The Viking Press.

Dunn, John M., 1990, *Interpreting Political Responsibility: Essays 1981–1989*, Princeton, NJ: Princeton University Press.

Dunn, John M., 2005, *Setting the People Free: The Story of Democracy*, London: Atlantic Books.

Dunn, John M., 2014, *Breaking Democracy's Spell*, New Haven, CT: Yale University Press.

Elliott, E.N. (ed.), 1860, *Cotton is King, and Pro-Slavery Arguments, Comprising the Writings of Hammond, Harper, Christy, Stringfellow, Hodge, Bledsoe, and Cartwright on this Important Subject*, Augusta, GA: Pritchard, Abbott & Loomis.

Ellis, Frank, 2000, *Rural Livelihoods and Diversity in Developing Countries*, Oxford: Oxford University Press.

Elwin, Verrier, 1964, *The Tribal World of Verrier Elwin: An Autobiography*, London & Bombay: Oxford University Press.

Engels, Frederick, 1975 [1845], 'The Condition of the Working Class in England – From Personal Observation and Authentic Sources', in *Karl Marx Frederick Engels Collected Works*, Volume 4, London: Lawrence & Wishart.

Engels, Frederick, 1977, 'The Magyar Struggle [1849]', in *Karl Marx Frederick Engels Collected Works*, Volume 8, London: Lawrence & Wishart.

Engels, Frederick, no date/a, *Herr Eugen Dühring's Revolution in Science [Anti-Dühring]*, London: Martin Lawrence Limited.

Engels, Frederick, no date/b, *Engels on Capital: Synopsis, Reviews, Letters and Supplementary Material*, London: Lawrence and Wishart.

Eribon, Didier, 2018, *Returning to Reims*, London: Penguin Books/Allen Lane.

Escobar, Arturo, 1995, *Encountering Development: The Making and Unmaking of the Third World*, Princeton, NJ: Princeton University Press.

Evans, Richard J., 2019, *Eric Hobsbawm: A Life in History*, London: Little, Brown.

Fanon, Franz, 1963, *The Wretched of the Earth*, New York: Grove Press.

Fasenfest, David, 2022, 'The Challenge for Sociology: The Value of the Critique', *Critical Sociology*, Vol. 48, No. 2.

Foenkinos, David, 2020, *The Mystery of Henri Pick*, London: Pushkin Press.

Fogel, Robert William, and Stanley L. Engerman, 1974, *Time on the Cross: Volume I – The Economics of American Negro Slavery*, London: Wildwood House.

Foucault, Michel, 1996, *Foucault Live: Collected Interviews, 1961–1984* (Edited by Sylvère Lotringer, Translated by Lysa Hochroch and John Johnston), New York, NY: Semiotext(e).

Fraser, Ronald, 1988, *1968: A Student Generation in Revolt*, London: Chatto & Windus.

Friedman, Milton, 1970, *The Counter-Revolution in Monetary Theory*, London: Institute of Economic Affairs.

Friedmann, Harriet, 2021, 'Origins of Peasant Studies', in Haroon Akram-Lodhi *et al.* (eds) *Handbook of Critical Agrarian Studies*.

Fröbel, Folker, Jürgen Heinrichs, and Otto Kreye, 1980, *The New International Division of Labour: Structural Unemployment in Industrialized Countries and Industrialization in Developing Countries*, Cambridge: Cambridge University Press.

Fukuyama, Francis, 1992, *The End of History and the Last Man*, London: Hamish Hamilton.

Fukuyama, Francis, 2006a, 'Guidelines for Future Nation-Builders', in Francis Fukuyama (ed.), *Nation-Building*, Baltimore, MD: The Johns Hopkins University Press.

Fukuyama, Francis, 2006b, *After the Neocons: America at the Crossroads*, London: Profile Books Ltd.

Fukuyama, Francis, 2018, *Identity: Contemporary Identity Politics and the Struggle for Recognition*, London: Profile Books.

Fukuyama, Francis, 2022, *Liberalism and Its Discontents*, London: Profile Books.

Galbraith, John Kenneth, 1999, *Name-Dropping: From FDR On*, Boston and New York: Houghton Mifflin Company.

Gerber, Julien-François, 2021, 'Rural Indebtedness', in Haroon Akram-Lodhi *et al.* (eds), *Handbook of Critical Agrarian Studies*.

Gibbon, Peter, and Michael Neocosmos, 1985, 'Some Problems in the Political Economy of "African Socialism"', in Henry Bernstein and Bonnie K. Campbell (eds.), *Contradictions of Accumulation in Africa: Studies in Economy and State*, London and Beverley Hills, CA: Sage Publications.

Giddens, Anthony (ed.), 2003, *The Progressive Manifesto: New Ideas for the Centre-Left*, Cambridge: Polity Press.

Giddens, Anthony, 1989, *Sociology*, Cambridge: Polity Press.

Giddens, Anthony, and Christopher Pierson, 1998, *Conversations with Anthony Giddens: Making Sense of Modernity*, Cambridge: Polity Press.

Glyn, Andrew, 2006, 'Will Marx be Proved Right?', *Oxonomics*, Vol. 1, No. 1.

Glyn, Andrew, 2007, *Capitalism Unleashed: Finance, Globalization, and Welfare*, Oxford: Oxford University Press.

Gould, Julius, 1977, *The Attack on Higher Education: Marxist and Radical Penetration*, London: Institute for the Study of Conflict.

Greco, Elisa, 2021, 'Qualitative Research', in Haroon Akram-Lodhi *et al.* (eds) *Handbook of Critical Agrarian Studies*.

Green, Peter, 1957, *The Sword of Pleasure*, London: John Murray.

Greene, Graham, 1970, *Collected Essays*, Harmondsworth: Penguin Books.

Grimshaw, Anna, and Keith Hart, 1994, 'Anthropology and the Crisis of the Intellectuals', *Critique of Anthropology*, Vol. 14, No. 3.

Guevara, Ernesto Che, 2022, *I Embrace You With All My Revolutionary Fervour: Letters 1947–1967*, London: Penguin Books.

Guha, Mechthild, 2014, *Danube, Ganges, and Other Life Streams*, Delhi: Permanent Black.

Guha, Ranajit (ed.), 1982–89, *Subaltern Studies I–VI*. New Delhi: Oxford University Press.

Guha, Ranajit, 1983, *Elementary Aspects of Peasant Insurgency in Colonial India*. Delhi: Oxford University Press.

Guha, Ranajit, 2010, *The Small Voice of History: Collected Essays* (Edited and with an Introduction by Partha Chatterjee), Delhi: Permanent Black.

Hall, Ruth, Jacobo Grajales, Ricardo Jacobs, Sérgio Sauer, Sheila Sheshia Galvin, and Annie Shattuck, 2023, 'Intertwined histories: *JPS* at 50, La Via Campesina at 30', *The Journal of Peasant Studies*, Vol. 50, No. 2.

Hall, Ruth, Marc Edelman, Saturnino Borras, Ian Scoones, Ben White, and Wendy Wolford, 2015, 'Resistance, Acquiescence or Incorporation? An Introduction to Land Grabbing and Political Reactions "From Below"', *The Journal of Peasant Studies*, Vol. 42, Nos. 3–4.

Hall, Stuart, 1960, 'The Supply of Demand', in E.P. Thompson (ed.), *Out of Apathy*, London: New Left Books.

Hall, Stuart, 1989, 'Ethnicity: Identity and Difference', *Radical America*, Vol. 23, No. 4.

Hardiman, David, 1995, 'Community, Patriarchy, Honour: Raghu Bhanagre's Revolt', *The Journal of Peasant Studies*, Vol. 23, No. 1.

Harris, Ralph (ed.), 1961, *Radical Reaction: Essays in Competition and Affluence*, London: Institute of Economic Affairs.

Harriss, John, 2021, 'Footloose Labour', in Haroon Akram-Lodhi *et al.* (eds), *Handbook of Critical Agrarian Studies*.

Hart, Keith, 1973, 'Informal Income Opportunities and Urban Employment in Ghana', *The Journal of Modern African Studies*, Vol. 11, No. 1.

Hart, Keith, 1982, *The Political Economy of West African Agriculture*, Cambridge: Cambridge University Press.

Hart, Keith, 1990a, 'Blacks in the World Economy', *Cambridge Anthropology*, Vol. 14, No. 2.

Hart, Keith, 1990b, 'Swimming into the Human Current', *Cambridge Anthropology*, Vol. 14, No. 3.

Hart, Keith, 1992, 'Market and State after the Cold War: The Informal Economy Reconsidered', pp. 214–227 in Roy Dilley (ed.), *Contesting Markets: Analyses of Ideology, Discourse and Practice,* Edinburgh: Edinburgh University Press.

Hart, Keith, 2000, 'Industrial Labour in India: The View from 19th Century Lancashire', *Critique of Anthropology*, Vol. 20, No. 4.

Hart, Keith, 2022, *Self in the World: Connecting Life's Extremes*, London and New York: Berghahn Books.

Hayek, F.A., 1973, *Economic Freedom and Representative Government*, London: Institute of Economic Affairs.

Hickel, Jason, 2022, *Less is More: How DeGrowth will Save the World*, London: Penguin Books.

Hobsbawm, Eric J., 1969, *Bandits,* London: Weidenfeld & Nicolson.

Hobsbawm, Eric J., 2000, *On the Edge of the New Century: In Conversation with Antonio Polito*, New York: The New Press.

Hobsbawm, Eric J., 2002, *Interesting Times: A Twentieth-Century Life*. London: Allen Lane.

Hochschild, Arlie Russell, 2016, *Strangers in Their Own Land*, New York: The New Press.

Houellebecq, Michel, 2022, *Interventions 2020*, Cambridge: Polity.

Howard, Anthony, 1964, 'We are the Masters Now – The General Election of 5 July 1945', in Michael Sissons and Philip French (eds.), *The Age of Austerity, 1945–51*, Harmondsworth: Penguin Books.

Hunold, Albert (ed.), 1961, *Freedom and Serfdom*, Dordrecht: D.Reidel Publishing Company.

Hutton, Will, and Anthony Giddens (eds.), 2000, *On the Edge: Living with Global Capitalism*, London: Jonathan Cape.

International Labour Office, 1924, *Unemployment in its National and International Aspects*, Geneva: ILO Studies and Reports, Series C (Unemployment) No. 9.

International Labour Office, 1972, *Employment, Incomes and Equality: A Strategy for Increasing Productive Employment in Kenya*, Geneva: International Labour Office.

Isenberg, Nancy, 2017, *White Trash: The 400-Year Untold History of Class in America*, London: Atlantic Books.

Jackson, Tim, 2021, *Post-Growth: Life after Capitalism*, Cambridge: Polity Press.

James, C.L.R., 1938, *The Black Jacobins*, London: Secker & Warburg.

Jan, Ali, and Barbara Harriss-White, 2021, 'Agricultural markets', in Haroon Akram-Lodhi *et al.* (eds) *Handbook of Critical Agrarian Studies*.

Jha, Praveen, and Paris Yeros, 2021, 'Crises of Capitalism in the Countryside: Debates from the South', in Haroon Akram-Lodhi *et al.* (eds) *Handbook of Critical Agrarian Studies*.

Joyce, Patrick, 1982, *Work, Society and Politics: The Culture of the Factory in Later Victorian England*, London: Methuen & Co., Ltd.

Joyce, Patrick, 1987, 'The Historical Meanings of Work: An Introduction', in Patrick Joyce (ed.), *The Historical Meanings of Work*, Cambridge: Cambridge University Press.

Joyce, Patrick, 1991, 'History and Post-modernism', *Past & Present*, No. 133.

Joyce, Patrick, 1994, *Visions of the People: Industrial England and the Question of Class, 1848–1914*, Cambridge: Cambridge University Press.

Joyce, Patrick, 1998, 'The Return of History: Postmodernism and the Politics of Academic History in Britain', *Past & Present*, No. 158.

Joyce, Patrick, 2010, 'What is the Social in Social History?', *Past & Present*, No. 206.

Joyce, Patrick, 2024, *Remembering Peasants: A Personal History of a Vanished World*, London: Allen Lane.

Kalecki, Michał, 1943, 'Political Aspects of Full Employment', *The Political Quarterly*, Vol. 14, No. 4.

Kautsky, Karl, 1984 [1895], 'The Competitive Capacity of Small-scale Enterprise in Agriculture', in Athar Hussain and Keith Tribe (eds.), *Paths of Development in Capitalist Agriculture*, London: Macmillan.

Kautsky, Karl, 1988 [1899], *The Agrarian Question* (2 volumes), London: Zwan Publications.

Kennedy, John F., 1964, *A Nation of Immigrants* (Introduction by Robert F. Kennedy), New York: Harper & Row, Publishers.

Kennedy, Stetson, 1946, *Southern Exposure*, New York: Doubleday & Co.

Khanna, Parg, 2022, *Move: How Mass Migration Will Reshape the World – and What It Means for You*, London: Weidenfeld & Nicolson.

Kothari, Ashish, Ariel Salah, Arturo Escobar, Federico Demaria, and Alberto Acosta (eds.), 2019, *Pluriverse: A Post-Development Dictionary*, New Delhi: Tulika Books.

Kumar, Arun, 2000, 'Beyond Muffled Murmurs of Dissent? Kisan Rumour in Colonial Bihar', *The Journal of Peasant Studies*, Vol. 28, No. 1.

Kundnani, Hans, 2023, *Eurowhiteness: Culture, Empire and Race in the European Project*, London: C. Hurst & Co. (Publishers), Ltd.

Laclau, Ernesto, 1990, *New Reflections on the Revolution of Our Time*, London: Verso.

Laclau, Ernesto, 2005a, 'Populism: What's in a Name?', in Francisco Panizza (ed.) *Populism and the Mirror of Democracy*, London: Verso.

Laclau, Ernesto, 2005b, *On Populist Reason*, London: Verso.

Laermans, Rudi, 2010, 'On Populist Politics and Parliamentary Paralysis: An Interview with Ernesto Laclau', *Open*, No. 20 (a special issue on *The Populist Imagination*).

Leacock, Stephen, 1916, *Essays and Literary Studies*, London: John Lane, The Bodley Head.

Leeds, Anthony, 1971, 'The Concept of "Culture of Poverty", Conceptual, Logical, and Empirical Problems with Perspectives from Brazil and Peru', pp. 9–37 in Eleanor Burke Leacock (ed.), *The Culture of Poverty: A Critique*, New York: Simon and Schuster.

Leeds, Anthony, and Elizabeth Leeds, 1970, 'Brazil and the Myth of Urban Rurality: Urban Experience, Work and Values in "Squatments" of Rio de Janeiro and Lima', pp. 229–285 in Arthur J. Field (ed.), *City and Country in the Third World: Issues in the Modernization of Latin America*, Cambridge, MA: Schenkman Publishing Company, Inc.

Leinius, Johanna, 2021, 'Pluriloguing Postcolonial Studies and Critical Agrarian Studies', in Haroon Akram-Lodhi *et al*. (eds) *Handbook of Critical Agrarian Studies*.

Lenin, V.I., 1964a [1899], 'The Development of Capitalism in Russia', *Collected Works*, Volume 3, Moscow: Foreign Languages Publishing House.

Lenin, V.I., 1964b [1899], 'Review: Karl Kautsky, *Die Agrarfrage*', *Collected Works*, Volume 4, Moscow: Foreign Languages Publishing House.

Lenin, V.I., 1964c [1899], 'Review: A. Bogdanov. A Short Course of Economic Science', *Collected Works*, Volume 4, Moscow: Foreign Languages Publishing House.

Levien, Michael, Michael Watts, and Yann Hairong, 2018, 'Agrarian Marxism', *The Journal of Peasant Studies*, Vol. 45. Nos. 5–6.

Lévi-Strauss, Claude, and Didier Eribon, 1991, *Conversations with Claude Lévi-Strauss* (Translated by Paula Wissing), London and Chicago, IL: The University of Chicago Press.

Li, Tania Murray, 2021, 'Foreword', in Haroon Akram-Lodhi *et al*. (eds) *Handbook of Critical Agrarian Studies*.

Lichtenstein, Alex, 1996, *Twice the Work of Free Labor: The Political Economy of Convict Labor in the New South*, London: Verso.

Liebknecht, Wilhelm, 1908, *Karl Marx: Biographical Memoirs* (Translated by Ernest Untermann), Chicago, IL: Charles H. Kerr & Company.

Littlejohn, Gary, 1973a, 'The Peasantry and the Russian Revolution', *Economy and Society*, Vol. 2, No. 1.

Littlejohn, Gary, 1973b, 'The Russian Peasantry: A Reply to Teodor Shanin', *Economy and Society*, Vol. 2, No. 3.

Littlejohn, Gary, 1977, 'Peasant Economy and Society', in Barry Hindess (ed.), *Sociological Theories of the Economy*, London: The Macmillan Press, Ltd.

Lucassen, Jan, 1993, 'Free and Unfree Labour before the the Twentieth Century: A Brief Overview', in Tom Brass, Marcel van der Linden, and Jan Lucassen, *Free and Unfree Labour*, Amsterdam: International Institute for Social History.

Lucassen, Jan, 2013, *Outlines of a History of Labour*, Amsterdam: IISH Research Paper 51.

Lucassen, Jan, 2022, *The Story of Work: A New History of Humankind*, New Haven, CT: Yale University Press.

Lundberg, George A., 1939, *Foundations of Sociology*, New York: The Macmillan Company.

MacAskill, William, 2015, *Doing Good Better: Effective Altruism and a Radical New Way to Make a Difference*, New York: Penguin/Random House.

Mackenzie Wallace, Donald, 1877, *Russia,* London: Cassell & Company, Limited.

Malefakis, Edward E., 1970, *Agrarian Reform and Peasant Revolution in Spain*, New Haven, CT: Yale University Press.

Malinowski, Bronislaw, 1967, *A Diary in the Strict Sense of the Term*, London: Routledge & Kegan Paul.

Mamet, David, 2023, *Everywhere an Oink Oink: An Embittered, Dyspeptic, and Accurate Report of Forty Years in Hollywood*, New York, NY: Simon & Schuster.

Marquardt, Felix, 2021, *The New Nomads: How the Migration Revolution is Making the World a Better Place*, London and New York: Simon & Schuster.

Marx, Karl, 1975, 'On the Jewish Question [1844]', in Karl Marx Frederick Engels *Collected Works*, Volume 3, London: Lawrence & Wishart.

Marx, Karl, 1976, *Capital Volume 1* (Introduced by Ernest Mandel, Translated by Ben Fowkes), Harmondsworth: Penguin Books.

Marx, Karl, 1977 [1848], 'Confessions of a Noble Soul', in Karl Marx Frederick Engels *Collected Works*, Volume 8, London: Lawrence & Wishart.

Marx, Karl, 1979 [1852], 'The Eighteenth Brumaire of Louis Bonaparte', in Karl Marx Frederick Engels, *Collected Works*, Volume 11, London: Lawrence and Wishart.

Marx, Karl, and Frederick Engels, 1934, *Correspondence 1846–1895*, London: Martin Lawrence Ltd.

Marx, Karl, and Frederick Engels, 1975, 'The Holy Family [1845]', in Karl Marx Frederick Engels *Collected Works*, Volume 4, London: Lawrence & Wishart.

Marx, Karl, and Frederick Engels, 1976 [1845–46], 'The German Ideology', in Karl Marx Frederick Engels *Collected Works*, Volume 5, London: Lawrence & Wishart.

Mattick, Paul, 1971, *Marx and Keynes: The Limits of the Mixed Economy*, London: Merlin Press.

McCallum, R.B., and Alison Readman, 1947, *The British General Election of 1945*, London: Oxford University Press.

McCarthy, Mary, 1953, *The Groves of Academe*, London: William Heinemann Ltd.

McCusker, Brent, Paul O'Keefe, Phil O'Keefe, Geoff O'Brian, 2013, 'Peasants, Pastoralists and Proletarians: Joining the Debates on Trajectories of Agrarian Change, Livelihoods and Land Use', *Human Geography*, Vol. 6, No. 3.

McKay, Ben, and Henry Veltmeyer, 2021, 'Industrial Agriculture and Agrarian Extractivism', in Haroon Akram-Lodhi *et al.* (eds) *Handbook of Critical Agrarian Studies*.

McKenzie, Robert T., and Allan Silver, 1968, *Angels in Marble: Working Class Conservatives in Urban England*, London: Heinemann Educational Publishers.

Mezzadra, Sandro, 2006, *Diritto di fuga. Migrationi, Cittadinanza, Globalizzazione*. Verona: ombre corte.

Mezzadra, Sandro, 2011a, 'How Many Histories of Labor? Towards a Theory of Postcolonial Capitalism', *Postcolonial Studies*, Vol. 14, No. 2.

Mezzadra, Sandro, 2011b, 'The Gaze of Autonomy: Capitalism, Migration, and Social Struggles', in Vicki Squire (ed.), *The Contested Politics of Mobility: Borderzones and Irregularity*, London: Routledge.

Mezzadra, Sandro, and Brett Neilson (eds.), 2013, *Border as Method, or the Multiplication of Labor*, Durham, NC: Duke University Press.

Miéville, China, 2022, *A Spectre, Haunting*, London: Head of Zeus Ltd.

Mishra, Pankaj, 2020, *Bland Fanatics: Liberals, Race and Empire*, London: Verso.

Mitrany, David, 1951, *Marx Against the Peasant*, Chapel Hill, NC: University of North Carolina Press.

Mouffe, Chantal, 2018, *For a Left Populism*, London: Verso.

Mudde, Cas, 2002, 'In the Name of the Peasantry, the Proletariat, and the People: Populisms in Eastern Europe', in Y. Mény and Y. Surel (eds.), *Democracy and the Populist Challenge*, London: Palgrave, Macmillan.

Munslow, Alun, and Robert A. Rosenstone (eds.), 2004, *Experiments in Rethinking History*, London: Routledge.

Murphy, Robert F., 1987, *The Body Silent*, New York: W.W.Norton.

Nanda, Meera, 2001, 'We Are All Hybrids Now: The Dangerous Epistemology of Postcolonial Populism', *The Journal of Peasant Studies*, Vol. 28, No. 2.

Nanda, Meera, 2004, *Prophets Facing Backward: Postmodern Critiques of Science and Hindu Nationalism in India*, New Brunswick: Rutgers University Press.

Newby, I.A. (ed.), 1968, *The Development of Segregationist Thought*, Homewood, IL: The Dorsey Press.

Nieboer, H.J., 1910, *Slavery as an Industrial System*, The Hague: Martinus Nijhoff.

Oakeshott, Michael, 1939, *The Social and Political Doctrines of Contemporary Europe* (with a Foreword by Ernest Barker), Cambridge: At the University Press.

Oakeshott, Michael, 1962, *Rationalism in Politics and Other Essays*, London: Methuen & Co. Ltd.

Oliveira, Gustavo, and Ben McKay, 2021, 'BRICS and Global Agrarian Transformations', in Haroon Akram-Lodhi *et al.* (eds) *Handbook of Critical Agrarian Studies*.

Patel, Sujata, 2008, *The Jan Breman Omnibus*, New Delhi: Oxford University Press.

Patnaik, Utsa, and Prabhat Patnaik, 2017, *A Theory of Imperialism*. New York: Columbia University Press.

Pattenden, Jonathan, 2021, 'Labour', in Haroon Akram-Lodhi *et al.* (eds) *Handbook of Critical Agrarian Studies*.

Pellew, Jill, and Miles Taylor (eds.), 2021, *Utopian Universities: A Global History of the New Campuses of the 1960s*, London & New York: Bloomsbury Academic.

Perkin, Harold, 1996, 'An Age of Great Cities', in Debra N. Mancoff and D.J. Trela (eds.), *Victorian Urban Settings: Essays on the Nineteenth-Century City and Its Contexts*, New York and London: Garland Publishing Inc.

Petras, James F., 1990, 'The Retreat of the Intellectuals', *Economic & Political Weekly*, Vol. 25, No. 38.

Petras, James F., 2002, 'A Rose by Any Other Name? The Fragrance of Imperialism', *The Journal of Peasant Studies*, Vol. 29, No. 2.

Powdermaker, Hortense, 1966, *Stranger and Friend: The Way of an Anthropologist*, New York: W.W. Norton & Company Inc.

Pradella, Lucia, and Rossana Cillo, 2021, 'Bordering the Surplus Population Across the Mediterranean: Imperialism and Unfree Labour in Libya and the Italian Countryside', *Geoforum*, Vol. 126.

Putzel, James, 2004, *The Politics of Participation: Civil Society, the State and Development Assistance* (Discussion Paper No. 1), London: Crisis States Development Research, Development Studies Institute, LSE.

Raper, Arthur F., 1936, *Preface to Peasantry: A Tale of Two Black Belt Counties*, Chapel Hill, NC: The University of North Carolina Press.

Raper, Arthur F., 1970 [1933], *The Tragedy of Lynching*, New York: Dover Publications, Inc.

Roberts, David, 1979, *Paternalism in Early Victorian England*, London: Croom Helm Ltd.

Rubin, Isaac Ilyich, 1979 [1929], *A History of Economic Thought* (Translated and Edited by Donald Filtzer), London: Ink Links Ltd.

Sanghera, Sathnam, 2021, *Empireland: How Imperialism Has Shaped Modern Britain*, London: Penguin/Viking.

Sanghera, Sathnam, 2024, *Empireworld: How British Imperialism Has Shaped the Globe*, London: Penguin/Viking.

Sathyamurthy, T.V., 1990, 'Indian Peasant Historiography: A Critical Perspective on Ranajit Guha's Work', *The Journal of Peasant Studies*, Vol. 18, No. 1.

Schlesinger, Rudolf, 1953, *Central European Democracy and its Background*, London: Routledge & Kegan Paul.

Scott, James C., 1976, *The Moral Economy of the Peasant: Rebellion and Subsistence in Southeast Asia*, New Haven, CT: Yale University Press.

Scott, James C., 1985, *Weapons of the Weak: Everyday Forms of Peasant Resistance*, New Haven, CT: Yale University Press.

Scott, James C., 2012, *Decoding Subaltern Politics: Ideology, Disguise, and Resistance in Agrarian Politics*, New Haven, CT: Yale University Press.

Seldon, Arthur (ed.), 1961, *Agenda for a Free Society*, London: Institute of Economic Affairs.

Seligman, Ben B., 1962, *Main Currents in Modern Economics: Economic Thought Since 1870*, New York: The Free Press.

Serge, Victor, 2004 [1925], 'The Impotence of the Intellectuals', in *Victor Serge: Collected Writings on Literature and Revolution* (Translated and Edited by Al Richardson), London: Francis Boutle Publishers.

Singh, Hira, 2002, 'Caste, Class and Peasant Agency in Subaltern Studies Discourse: Revisionist Historiography, Elite Ideology', *The Journal of Peasant Studies*, Vol. 30, No. 1.

Sinha, Indradeep, 1982, *Some Questions Concerning Marxism and the Peasantry*, New Delhi: Communist Party of India.

Skinner, Quentin (ed.), 1985, *The Return of Grand Theory in the Human Sciences*, Cambridge: Cambridge University Press.

Smith, Horace, 1890 [1836], *The Tin Trumpet*, London: George Routledge and Sons, Limited, Broadway, Ludgate Hill.

Soper, Kate, 2020, *Post-Growth Living*, London: Verso.

Sorokin, Pitrim, and Carle C. Zimmerman, 1939, *Principles of Rural-Urban Sociology*. New York: Henry Holt and Company.

Stallabrass, Julian, 2006, *High Art Lite: The Rise and Fall of Young British Art*, London: Verso.

Steinbeck, John, 1939, *The Grapes of Wrath*, London: William Heinemann Ltd.

Steinfeld, Robert J., 2001, *Coercion, Contract, and Free Labor in the Nineteenth Century*, Cambridge: Cambridge University Press.

Stepniak, Sergius, 1905, *The Russian Peasantry: Their Agrarian Condition, Social Life and Religion*, London: George Routledge & Sons, Limited.

Stiglitz, Joseph E., 2024, *The Road to Freedom: Economics and the Good Society*, London: Allen Lane.

Stowe, Harriet Beecher, c. 1860, *The Key to Uncle Tom's Cabin; Presenting the Original Facts and Documents upon which the Story is Founded*, London: Clarke, Beeton, and Co., 148, Fleet Street.

Streek, Wolfgang, 2017, *Between Charity and Justice: Remarks on the Social Construction of Immigration Policy in Rich Democracies*. Danish Centre for Welfare Studies, University of Southern Denmark, DaWS Working Paper – 5.

Sumner, Andy, and Meera Tiwari, 2009, *After 2015: International Development Policy at a Crossroads*, Basingstoke: Palgrave Macmillan.

Sutch, Richard, 1975, 'The Treatment Received by American Slaves: A Critical Review of the Evidence Presented in *Time on the Cross*'. *Explorations in Economic History*, Vol. 12, No. 4.

Sweezy, Paul M., 1946, *The Theory of Capitalist Development* (With a Forword by Maurice Dobb), London: Denis Dobson Limited.

Taylor, Laurie, et al. (2004) *How to Get Promoted: A Career Guide for Academics*, London: THES pamphlet.

Thaler, Richard, and Cass Sunstein, 2008, *Nudge: Improving Decisions about Health, Wealth, and Happiness*, New Haven, CT: Yale University Press.

Tharoor, Shashi, 2017, *Inglorious Empire: What the British Did to India*, London: C. Hurst & Co. (Publishers) Ltd.

Thomas, William I., and Florian Znaniecki, 1927, *The Polish Peasant in Europe and America*, Volume 1, New York: Alfred A. Knopf.

Thompson, E.P., 1960, 'The Point of Production', *New Left Review*, No. 1 (January/February).

Thorner, Daniel, and Alice Thorner, 1962, *Land and Labour in India*, New Delhi: Asia Publishing House.

Tolnay, Stewart E., and E.M. Beck, 1995, *A Festival of Violence: An Analysis of Southern Lynchings, 1882–1930*, Urbana and Chicago, IL: University of Illinois Press.

Tolstoy, Leo, 1905, 'The School at Yásnaya Polyána', in *Pedagogical Articles* (edited and translated by Leo Wiener), London: G.J. Howell & Co., 32 Newgate Street, E.C.

Trotsky, Leon, 1934, *The History of the Russian Revolution* (Translated by Max Eastman). London: Victor Gollancz Ltd.

Trotsky, Leon, 1962, *The Permanent Revolution [1928] and Results and Prospects [1906]*, London: New Park Publications Limited.

Trotsky, Leon, 1967 [1936], *The Revolution Betrayed*, London: New Park Publications.

UK Parliamentary Committee, 1932, *Report on Ministers' Powers, Presented by the Lord High Chancellor to Parliament by Command of His Majesty, April 1932*, London: HMSO.

van der Linden, Marcel, 2023, *The World Wide Web of Work: A History in the Making*. London: UCL Press.

van der Linden, Marcel, and Jan Breman, 2020, 'The Return of Merchant Capital', *Global Labour Journal*, Vol. 11, No. 2.

van der Linden, Marcel, and Karl-Heinz Roth, 2018, 'Marxism' or Marx's Method? A Brief Response to Brass', *Science & Society*, Vol. 82, No. 1.

van der Ploeg, Jan Douwe, 2021, 'Peasants', in Haroon Akram-Lodhi *et al.* (eds) *Handbook of Critical Agrarian Studies*.

Veltmeyer, Henry, 2021, 'The Interface of critical development studies and critical agrarian studies', in Haroon Akram-Lodhi *et al.* (eds.) *Handbook of Critical Agrarian Studies*.

Venturi, Franco, 1960, *Roots of Revolution*, London: Weidenfeld & Nicolson.

von Fürer-Haimendorf, Christof, 1990, *Life Among Indian Tribes: The Autobiography of an Anthropologist*, Delhi: Oxford University Press.

von Schlegel, Frederick, 1848, *The Philosophy of History, in a Course of Lectures Delivered at Vienna* (Translated from the German by James Burton Robertson, Esq.), London: Henry G. Bohn, York Street, Covent Garden.

Walters, Everett, 1960, 'Populism: Its Significance in American History', in Donald Sheehan and Harold C. Syrett (eds.), *Essays in American Historiography: Papers Presented in Honor of Allan Nevins*, New York: Columbia University Press. Pp. 217–30.

Warriner, Doreen, 1939, *Economics of Peasant Farming*, London: Oxford University Press.

Warriner, Doreen, 1948, *Land and Poverty in the Middle East*, London and New York: Royal Institute of International Affairs.

Warriner, Doreen, 1950, *Revolution in Eastern Europe*, London: Turnstile Press.

Warriner, Doreen, 1957, *Land Reform and Development in the Middle East*, London and New York: Royal Institute of International Affairs.

Warriner, Doreen, 1969, *Land Reform in Principle and Practice*, Oxford: Clarendon Press.

Warriner, Henry, 2021, *Doreen Warriner's War: Refugees from Prague, Food for Egypt, Starvation in Belgrade*, London: The Book Guild, Ltd.

Watts, Michael J., 2021, 'The Agrarian Question', in Haroon Akram-Lodhi *et al.* (eds) *Handbook of Critical Agrarian Studies*.

Whiteside, Andrew Gladding, 1962, *Austrian National Socialism before 1918*, The Hague: Martinus Nijhoff.

Whiteside, Andrew Gladding, 1975, *The Socialism of Fools: Georg Ritter von Schönerer and Austrian Pan-Germanism*, Berkeley, CA: University of California Press.

Widgery, David (ed.), 1976, *The Left in Britain 1956–68*. Harmondsworth: Penguin Books.

Williams, Eric, 1964, *Capitalism and Slavery*, London: Andre Deutsch Limited.

Williams, Joan C., 2017, *White Working Class*, Boston, MA: Harvard Business Review Press.

Woofter, T.J., 1936, *Landlord and Tenant on the Cotton Plantation*, Washington, DC: Division of Social Research, Works Progress Administration.

Worsley, Peter, 1972, 'Fanon and the "Lumpenproletariat"', in Ralph Miliband and John Saville (eds), *The Socialist Register 1972*, London: The Merlin Press Ltd.

Worsley, Peter, 1984, *The Three Worlds: Culture and World Development*, London: Weidenfeld & Nicolson.

Worsley, Peter, 2008, *An Academic Skating on Thin Ice*, New York: Berghahn Books.

Wortman, Richard, 1967, *The Crisis of Russian Populism*, London: Cambridge University Press.

Zimmerman, J.G., 1800, *Aphorisms and Reflections on Men, Morals and Things*, London: Thomas Maiden, Sherbourne Lane.

Žižek, Slavoj, 2014, *Trouble in Paradise: From the End of History to the End of Capitalism*, London: Penguin Books.

Žižek, Slavoj, 2023, *Too Late to Awaken: What Lies Ahead When There is No Future*, London: Allen Lane/Penguin Books.

Author Index

Acosta, A. 10–11
Adorno, T.W. 227
Akram-Lodhi, H. 97, 103, 105, 111–112, 113–114, 116
Amin, S. 12
Amis, K. 86
Angelo, L. 48
Arnold, D. 12
Augstein, H.F. 26
Ayers, E.L. 46

Balibar, E. 72
Banaji, J. 97, 126–27, 132, 135, 222
Barran, D.H. 7
Basso, P. 67ff., 71–72, 74ff., 80–81, 83
Bauer, A.J. 136
Bayly, C.A. 12
Beck, E.M. 46
Beck, U. 16
Benn, T. 161
Bernstein, H. 67, 111 *passim*, 186, 207
Berry, S. 107
Beveridge, W.H. 5, 18, 26, 53, 59ff., 72, 80, 81–82, 242
Beverley, J. 12
Bhadra, G. 12
Bhopal, K. 31
Blackburn, R. 87
Blok, A. 208
Bodley, J.H. 3
Borges, J.L. 161
Borras, S. 106, 112, 114ff
Bradbury, M. 19, 85 *passim*
Bradley, G.M. 67, 69 *passim*, 81, 83
Brass, T. 4, 12, 15, 17, 26ff., 33, 38, 40–41, 47, 49, 53, 67, 69, 72, 87, 90, 97–97, 101ff., 107–108, 111, 113ff., 122, 124–25, 127, 129–30, 135–36, 144–45, 152, 155, 161, 171, 173, 178, 182–83, 185, 186–87, 190, 192, 194–95, 201–202, 204, 205, 208, 213, 216, 220ff., 227ff., 234, 237
Breman, J. 4, 17, 69, 97, 107, 126, 135, 173, 182–83, 185 *passim*, 196–97
Bruno, S. 66
Burgess, A. 89
Byres, T.J. 4, 111ff., 116

Campling, L. 116
Cannadine, D. 224ff., 230ff., 236, 238
Caponigri, A.R. 149
Capps, G. 116
Chaturvedi, V. 216
Chayanov, A.V. 109, 175, 206–207, 212ff
Chesterton, C. 40
Christian, C. 46
Cillo, R. 74
Clark, A. 161
Cockburn, A. 87
Cohn-Bendit, D. 88
Cohn-Bendit, G. 88
Collins, J. 67
Cox, C.B. 86
Croce, B. 149
Croll, A. 202
Cromer, Earl of 7
Crossman, R.H.S. 6, 161
Curry, P. 224, 229ff., 236ff
Curtis, B. 10

Daniel, P. 48
Das, R.J. 114–15, 219
de Janvry, A. 108
de Noronha, L. 67, 69 *passim*, 81, 83
Deas, M. 134
Demaria, F. 10–11
Desai, V. 3
Dietz, K. 97, 103, 105, 108, 113, 116
Dixon, T. 34 *passim*, 49ff
Dobb, M. 18, 59, 63ff., 81, 91
Drage, G. 213
Draper, T. 9–10
Dunn, J.M. 149
Dyson, A.E. 86

Edelman, M. 106, 114–15, 207, 217
Elliott, E.N. 39
Ellis, F. 108
Elwin, V. 163
Engels, B. 97, 103, 105, 111–112, 113–114, 116
Engels, F. 26–27, 30, 49, 55ff., 76, 81, 123, 137 *passim*, 173–74, 207, 211, 225–26, 233
Engerman, S.L. 103, 124–25, 128
Eribon, D. 162–63, 210

AUTHOR INDEX

Escobar, A. 10–11, 145
Evans, R.J. 98

Fanon, F. 138, 172
Fasenfest, D. 219
Foenkinos, D. 52
Fogel, R.W. 103, 124–25, 128
Foucault, M. 7, 128, 153, 207
Fraser, R. 87
Friedman, M. 7, 190
Friedmann, H. 112, 116
Fröbel, F. 124
Fukuyama, F. 4, 17, 20, 53, 147 *passim*, 155 *passim*, 243

Galbraith, J.K. 165
Galvin, S.S. 116–117
Gerber, J-F. 108
Gibbon, P. 111
Giddens, A. 16, 86, 160, 162
Glyn, A. 26, 59, 65, 81
Gould, J. 86
Grajales, J. 116–17
Greco, E. 112
Green, P. 151
Greene, G. 134
Grimshaw, A. 169
Guevara, E. 224
Guha, M. 12–13
Guha, R. 11ff., 90, 145

Hall, R. 106, 116–17
Hall, S. 7, 138, 231
Hardiman, D. 12
Harris, R. 7
Harriss, J. 107
Harriss-White, B. 112
Hart, K. 17, 20, 163 *passim*, 171 *passim*, 180–81, 205, 243–44
Hayek, F.A. 152
Heinrichs, J. 124
Hickel, J. 10
Hobsbawm, E.J. 7, 98, 124, 160, 162, 208, 229
Hochschild, A.R. 27
Houellebecq, M. 9
Howard, A. 5
Hunold, A. 7
Hutton, W. 16

International Labour Office 60, 170
Isenberg, N. 27

Jackson, T. 10
Jacobs, R. 116–17
James, C.L.R. 124, 166, 172, 194
Jan, A. 112
Jha, P. 112, 114
Johnson, H.G. 7
Joyce, P. 21, 201 *passim*, 211 *passim*, 243–44

Kalecki, M. 18, 59, 63–64, 81
Kautsky, K. 109ff., 112, 118, 142, 207
Kay, C. 111–12
Kennedy, J.F. 68
Kennedy, S. 48
Khanna, P. 68
Kothari, A. 10–11
Kreye, O. 124
Kumar, A. 12
Kundnani, H. 31ff

Laclau, E. 21, 108, 224, 226 *passim*, 238
Laermans, R. 228
Leacock, S. 121
Leeds, A. 169–70
Leeds, E. 169–70
Liebknecht, W. 223–24
Leinius, J. 108
Lenin, V.I. 10, 26, 28, 30, 45, 49, 75, 104, 107, 109ff., 111–12, 118, 124, 137, 142, 173, 207, 211, 213–14, 216–17, 221, 224–25, 233
Lévi-Strauss, C. 7, 162, 210
Li, T.M. 106
Lichtenstein, A. 48
Littlejohn, G. 208
Lucassen, J. 8, 19, 122, 129ff., 133, 141, 222
Lundberg, G.A. 86

MacAskill, W. 16
Mackenzie Wallace, D. 213
Malefakis, E.E. 206
Malinowski, B. 163
Mamet, D. 1, 95
Marquardt, F. 68
Marx, K. 6, 18, 26ff., 49, 53ff., 58, 64, 66ff., 74ff., 81, 88, 93, 109, 123–24, 129, 137, 139–40, 142, 153, 162, 172ff., 186, 194–95, 207, 211, 215, 221 *passim*, 233–34, 241

Mattick, P.   6
McCallum, R.B.   6, 63
McCarthy, M.   92
McCusker, B.   183
McKay, B.M.   97, 103, 105, 108, 112–13, 116
McKenzie, R.T.   203
Mezzadra, S.   67, 71–72, 74, 83
Miéville, C.   221, 225
Mishra, P.   31
Mitrany, D.   108
Mouffe, C.   21, 224, 226, 228 *passim*, 238
Mudde, C.   228
Munslow, A.   33–34
Murphy, R.F.   163

Nanda, M.   108, 114–15
Neilson, B.   72
Neocosmos, M.   111
Newby, I.A.   26
Nieboer, H.J.   186–87

O'Brian, G.   183
O'Keefe, Paul   183
O'Keefe, Phil   183
Oakeshott, M.   149
Oliviera, G.   108

Patel, S.   193
Patnaik, P.   144
Patnaik, U.   4, 97, 144, 185, 194
Pattenden, J.   107, 113
Pellew, J.   84
Perkin, H.   202
Petras, J.F.   108, 114–15, 124
Pierson, C.   162
Potter, R.B.   3
Powdermaker, H.   163
Pradella, L.   74
Putzel, J.   7

Raper, A.F.   35, 39, 42 *passim*
Readman, A.   6, 63
Roberts, D.   203
Rosenstone, R.A.   33–34
Roth, K-H.   122
Rowland, G.   7
Rubin, I.I.   2

Sadoulet, E.   108

Salah, A.   10–11
Samaddar, R.   72
Sanghera, S.   31
Sathyamurthy, T.V.   12
Sauer, S.   116–17
Schlesinger, R.   6
Scoones, I.   106, 114–15
Scott, J.C.   107, 114–15, 145, 207, 216
Seldon, A.   7
Seligman, B.B.   2
Serge, V.   84
Shattuck, A.   116–17
Silver, A.   203
Singh, H.   12
Sinha, I.   12
Skinner, Q.   7
Smith, H.   219
Soper, K.   10
Sorokin, P.   108
Stallabrass, J.   17
Steinbeck, J.   52
Steinfeld, R.J.   127, 129
Stepniak, S.   213
Stiglitz, J.E.   147
Stowe, H.B.   37
Streek, W.   69
Sumner, A.   3
Sunstein, C.   16
Sutch, R.   103, 125
Sweezy, P.M.   18, 58–59, 63–64, 81

Taylor, L.   95
Taylor, M.   84
Thaler, R.   16
Tharoor, S.   31
Thomas, W.I.   205
Thompson, E.P.   6, 91, 123, 230–31
Thorner, A.   8, 165
Thorner, D.   8, 191
Tiwari, M.   3
Tolnay, S.E.   46
Tolstoy, Count L.   102
Trotsky, L.   4, 6, 26, 28, 49, 54, 110–11, 124, 142, 207, 213, 216, 221, 224

UK Parliamentary Committee   143

van der Linden, M.   8, 17, 19, 122 *passim*, 133 *passim*, 141, 205, 243–44

van der Ploeg, J.D.   107, 113–14, 116, 207, 217
Veltmeyer, H.   112, 116
Venturi, F.   108
von Fürer-Haimendorf, C.   163
von Schlegel, F.   25

Walters, E.   9–10
Warriner, D.   162
Warriner, H.   162
Watts, M.J.   103, 109ff., 113–14
White, B.   106, 116
Whiteside, A.G.   144
Widgery, D.   87
Williams, E.   124, 194

Williams, J.C.   27
Willms, J.   16
Wolford, W.   106
Woofter, T.J.   43, 44
Worsley, P.   124, 138, 163
Wortman, R.   108

Yeros, P.   112ff

Zimmerman, C.C.   108
Zimmerman, J.G.   101
Žižek, S.   14, 128
Znaniecki, F.   205

Subject Index

academia   8, 16–17, 20–21, 86, 89, 91, 97ff., 146, 159, 163, 165, 167ff., 220–21, 223, 225, 233, 239, 244, 247
   Cambridge   64, 87, 134, 148, 164 *passim*, 231
   Chicago   164–65, 167
   East Anglia   86, 165
   employment in   20, 91, 96, 99–100, 163 *passim*, 180
   Essex   87
   fashion and   XI, 9, 10, 16, 19, 85, 90, 92 *passim*, 102, 118, 122, 135, 156, 163, 220–21, 223, 224, 238, 242
   journals   94ff., 101ff., 116ff., 127, 185, 207, 221–22, 226, 242
   London School of Economics   166
   Manchester   165, 180
   Oxford   87–88, 231
   polemics in   187, 219, 220–21, 225
   Pretoria   172
   publishing in   19, 84–85, 94, 96, 98, 115, 118, 242
   Sussex   3, 88, 170
   Washington   165
Africa
   Angola   166
   Ghana   20, 168ff., 176, 178
   Kenya   170
   Rhodesia   3
   South Africa   143, 232
agrarian reform   5, 39, 40, 148, 150, 158. *See also* land tenure
Asia
   China   57, 65
   Japan   128, 158
   Korea   158
   Taiwan   158

Beveridge, William   5, 18, 26, 53, 59ff., 72, 80ff., 242
   'four giant evils'   60
   1942 Report   5, 61–62
borders   18, 68ff., 76, 78, 80, 82–83, 179, 241–42. *See also* capitalism, citizenship, labour market competition, migrants, reserve army

'right to escape'   71
   control of   62, 68, 70, 73, 77ff., 178
   open   18, 32, 68ff., 75 *passim*, 105, 157–58, 178, 234, 241–42

capitalism. *See also* class, labour, Marx, nationalism, production relations, reserve army
   'nicer'/'caring'   4, 10, 21, 69, 72, 107, 122, 131, 142, 146, 190, 243
   deregulation   6–7, 18, 65–66, 130, 151–52, 156, 176, 179, 184
   financial   28, 35, 104, 142
   globalization   16, 18, 53, 82, 183, 244
   industrial   128, 153, 194
   laissez-faire   2, 4ff., 20, 53, 62, 65–66, 69, 78, 83, 146, 151ff., 190, 228, 233, 243
   outsourcing   15, 65, 80, 124, 157
   privatisation   6, 66, 151–52, 156, 179
   restructuring   15, 42, 59, 66, 75, 130, 133, 150, 157, 184
Chayanov, A.V.   109, 175, 206–207, 212ff
citizenship   4, 39, 70, 72–73, 157ff., 177, 247
class
   bourgeoisie   6, 12, 16, 20, 45, 57, 67, 89, 94, 99, 108, 110, 112, 137, 139–40, 144, 146–47, 158, 174–75, 202, 220–21, 223–24, 225, 233
   consciousness   7–8, 12, 26, 55–56, 64, 77, 106, 108, 121, 136, 139, 172, 178, 182, 202–203, 210, 217, 222, 226–27, 234, 236, 241, 243
   depeasantisation   105, 107–108
   formation   2, 104, 108, 123, 136, 140, 183, 217, 234
   landlord   12, 42, 44–45, 46ff., 103, 134, 185, 188, 191, 205, 208, 233
   lumpenproletariat   19, 56, 123, 128, 136ff., 142, 174, 233, 243
   oligarch   151
   petite bourgeoisie   12, 110, 174, 233
   planter   38 *passim*, 48, 246
   struggle   5–6, 14, 19, 27, 35, 53ff., 67, 74ff., 78, 107, 115, 129ff., 133, 136, 138, 140–41, 154, 156, 176, 184, 186, 194, 203, 232ff., 236, 240–41

colonialism   12–13, 18, 70, 144, 216, 247
culture. *See also* historiography, populism, postmodernism
 'culture of poverty'   169
 hegemony   16, 21, 39, 89, 106, 110, 117, 228ff., 233ff., 242
 tradition   37, 88, 102, 154–55, 211, 217, 228, 237

democracy   20, 61, 143, 146ff., 151, 156ff., 179, 228, 243
development
 de-growth   14
 economic   XI, 15, 33, 42, 81, 90, 105, 130, 140, 146, 240
 grand narratives   21, 202–203, 237
 informal sector and   17, 20, 169ff., 175–76, 243
 macro-level   7, 105–106
 methodology   8, 106, 125, 129, 134, 168, 204ff., 217, 236, 243
 micro-level   2, 7–8, 16, 105
 modernity and   7ff., 11, 13, 21, 30, 33–34, 72, 88, 121, 129, 132, 142, 155, 202, 209, 211, 217, 237ff
 NGOs   3, 15, 72, 82, 172
 post-development   9ff., 14, 118
 productivism   10–11
 progress and   2, 4–5, 9ff., 58, 62, 104, 147, 155–56, 194, 211, 217, 242
 underdevelopment   175
 urbanization   11, 14

Engels, Frederick   26–27, 30, 49, 55ff., 76, 81, 123, 137 *passim*, 173–74, 207, 211, 225–26, 233
environment   XI, 10, 14ff., 104
 climate change   XI, 1, 4, 14–15, 104
Europe   14, 16, 32, 65, 67, 69, 75, 78, 86, 89, 90, 95, 128, 144–45, 157, 162, 177, 206, 228
 Austria   1, 12, 57, 143–44
 Croatia   91–92
 Czechoslovakia   162, 144
 Eastern   13, 65, 75, 89, 110, 141, 157, 162, 228
 Germany   28, 37, 68, 69, 144
 Hungary   30
 Ireland   57, 76–77, 81, 139–40, 155, 205, 206
 Italy   57, 66, 68, 155, 206

 Netherlands   57, 187, 228
 Poland   206, 215
 Spain   134, 206
 Yugoslavia   91

feudalism   152–53, 158, 193
 semi-feudalism   4, 112, 133, 144, 187, 197
film   1, 34, 39, 49, 95, 151, 162, 179, 201, 232, 247
film directors
 Burge, Stuart   247
 Curtis, Adam   179
 Griffith, D.W.   34
 Howard, Leslie   162
 Knoles, Harry   39
 Spielberg, Steven   162, 201
 Stanton, Andrew   151
 Tati, Jacques   1
 Welles, Orson   247
films
 Birth of a Nation (1915)   34, 49
 Bolshevism on Trial (1911)   39
 Finding Nemo (2003)   151
 Indiana Jones (1981–2008)   162, 201
 Jour de fête (1949)   1
 Les Vacances de Monsieur Hulot (1953)   1
 Othello (1965)   247
 Pimpernel Smith (1941)   162
 Russia 1985–1999: TraumaZone (2022)   179
 The Tragedy of Othello (1951)   247

Gandhi, M.K.   11, 172
gender   3, 36, 48, 53, 58, 66, 77, 90, 202, 226, 236, 240–41, 247
Guha, Ranajit   11ff., 90, 145

health   5, 7, 20, 56, 63, 70, 80, 163–64, 178, 180
 AIDS   98
historiography   11–12, 25, 34, 90, 103–104, 121, 125, 141, 145, 162, 217, 229ff., 236, 244
 autobiography and   20, 40, 161ff., 180, 204
 cliometric   103, 125, 128, 129, 132, 141
 folklore   21, 201, 208–209, 217
 global labour history   8, 17, 19, 102, 122, 124ff., 129, 133–34, 141, 205, 243–44
 social history   21, 33, 93, 201, 206, 229ff., 236, 243
 whig   133

SUBJECT INDEX 275

human rights 2, 4, 18, 54, 68, 72–73, 177, 179, 108, 241
  asylum 69, 80
  reparations as 32, 54, 68

identity. *See also* nationalism, postmodernism
  Anglo-Saxon 35, 41–42
  customs and 10, 30, 150, 242
  Eurocentric xi, 9, 13–14, 21, 90, 102, 145, 194, 237
  hyphenated 155
  Judaism 27ff., 57, 62, 82, 162
  kinship 21, 37, 72, 185
  rural 9, 11, 13ff., 19, 21, 33, 40ff., 46, 48, 51, 88, 89, 101, 102ff., 107, 109, 110, 118, 124, 133–34, 136ff., 142, 145, 169–70, 175, 194, 204ff., 208, 215ff., 237, 242–43
  tradition and 9, 13, 27, 33, 35, 37, 44, 51, 88, 102, 110, 118, 129, 150, 154–55, 172, 185, 193, 206, 211, 217, 228, 237, 242
imperialism 4, 31, 138, 143–44, 173, 244
  Empire and 31–32, 102, 167, 215, 242, 245
India 8, 11ff., 15, 65, 97, 103, 126, 136, 144, 186–87, 194, 208, 216, 233
  Bardoli 192–93
  Communist Party of India (CPI) 12–13
  Gujarat 186–87, 191, 193–94, 197
  Haryana 195
  Tata 15

James, C.L.R. 124, 166, 172, 194

Keynes, J.M. 2, 59, 62–63, 162

labour
  attached 8, 73, 125, 188–89, 192
  bonded 17, 20, 21, 72, 107, 126, 129, 132, 135–36, 182 *passim*, 191 *passim*, 208, 243
  casual 47, 60, 66, 72, 184–85, 193, 196
  child 66, 67
  commodified 48, 126–27, 130, 135–36
  coolie 185, 187
  deskilling 57, 90
  division of 33, 66, 124, 168, 214, 228
  female 47, 62, 66–67, 82, 89, 125, 184ff
  hali 187ff., 192, 196
  indentured 31, 125–26, 132
  permanent 56, 64, 66, 71, 91, 145, 157, 164, 184, 193, 196
  seasonal 51, 62, 184–85
labour market competition 4, 17–18, 25ff., 28, 31ff., 35, 39, 44, 49–50, 52, 57, 60, 63, 71, 73, 78 *passim*, 90, 138–39, 141, 146, 150, 157ff., 178–79, 183, 204, 227, 239, 241–42, 245ff
labour process 15, 55, 57ff., 65ff., 112, 124, 130, 133, 182, 184–85, 246
  decomposition/recomposition 66, 130, 184
labour-power. *See* capitalism, development, labour market competition, Marx
land tenure. *See also* landlord, peasant, sharecropper, tenant
  collective 2, 108, 150
  cooperative 150
  estate 40, 109, 134, 136, 158–59
Latin America 124, 126, 128, 208, 216, 228, 233
  Argentina 224, 232
  Bolivia 216
  Brazil 232
  Chile 136, 158, 232
  Colombia 134
  Cuba 224
  Guatemala 158
  Mexico 65, 134
  Peru 136, 216
  Uruguay 232
liberalism 17, 20, 55, 59, 82, 143 *passim*, 153 *passim*, 176, 180, 243
  classical 17, 20, 147ff., 151ff., 180, 243
  individualism 16, 90, 149, 151, 174–75, 179, 180
  neoliberalism 4, 7, 9, 18, 20, 66, 94, 142, 144–45, 147–48, 150ff., 154ff., 159, 176, 178ff., 190, 202, 222, 228, 234–35

Marx, Karl 6, 18, 26ff., 49, 53ff., 58, 64, 66ff., 74ff., 81, 88, 93, 109, 123–24, 129, 137, 139–40, 142, 153, 162, 172ff., 186, 194–95, 207, 211, 215, 221 *passim*, 233–34, 241
Marxism. *See also* Frederick Engels
  agrarian question 104ff., 109 *passim*
  Kautsky, Karl 109ff., 118, 142, 207
  Lenin, V.I. 10, 26, 28, 30, 45, 49, 75, 104, 107, 109ff., 111–12, 118, 124, 137, 142, 173, 207, 211, 213–14, 216–17, 221, 224–25, 233
  Luxemburg, Rosa 124, 207

Marxism (*cont.*)
  revolutionary   3, 8, 18, 26, 73, 87ff., 91ff.,
    99, 138, 145, 202, 215–16, 221, 224ff
  Trotsky, Leon   4, 6, 26, 28, 49, 54, 110–111,
    124, 142, 207, 213, 216, 221, 224, 226
middle east
  Afghanistan   149
  Iraq   149
migrants   52–53, 58–59, 66–67, 69, 72
  passim, 80ff., 136, 138ff., 144, 157–58,
  169, 193, 196, 241. *See also* capitalism,
  identity, reserve army
  refugees   80, 162
mode of production   2, 20, 33, 66, 122,
  132, 186, 191–92, 194–95, 229. *See also*
  capitalism, feudalism, socialism

nationalism   1, 12–13, 26, 69–70, 73, 83, 143,
  146–47, 155, 201, 204, 216, 227, 236,
  241, 247
  patriotism   154, 202, 208, 236

peasant. *See also* capitalism, Chayanov,
    identity, populism, postmodernism
  differentiation   21, 45, 109, 213–14
  family   51, 104, 175, 213–14
  middle   110, 211
  poor   5, 136, 214, 232–33, 241
  rich   12, 113, 130, 136, 184, 212, 214, 233
  smallholding   21, 51, 42–43, 74, 101, 105,
    107, 110, 111, 113, 115, 118, 212, 215–16, 242
  Via Campesina   117
politics
  communism   92, 148, 179, 207, 221,
    225, 232
  conservatism   3, 6, 8, 9, 17, 19, 31, 63, 78ff.,
    87, 96, 102, 108, 112, 115, 141, 146, 150,
    155, 173, 194, 202ff., 207, 215–16, 219, 221,
    230, 236–37
  fascism   83, 162, 207
  humanism   90, 171–72, 174, 176–77
  Nazis   69, 155, 207, 227
  socialism   XI, 4, 5, 7 *passim*, 18, 20–21, 26,
    39, 51, 60, 74, 82ff., 90, 99, 102, 104, 110,
    112, 122, 131, 138, 143 *passim*, 153, 173, 176,
    180, 190, 207, 221, 223–24, 234, 243–44
populism. *See also* identity, nationalism,
    peasant, postmodernism
  'ethno-populism'   228

'farmer first'   15
'food regimes'   108
'food security'   108, 189
'food sovereignty'   108
agrarian myth   17, 40, 50–51, 88, 108, 118,
  217, 240
anti-capitalism   8–9, 19, 38, 50, 76, 102,
  142, 203, 245
Critical Agrarian Studies   19, 97, 101, 103,
  105ff., 111, 112, 114–15, 117–18, 242–43
instinctivism   227
new social movements   10, 102, 234–35
postmodernism 'floating signifier'   233, 235
  Subaltern Studies   11ff., 54, 74, 90, 102,
    108, 145, 155, 204, 216
  alterity   21, 203
  aporia   7, 16, 21, 34, 122, 132, 149, 162, 173–
    74, 203, 210, 221, 229
  difference   21, 30, 34–35, 39, 203, 237, 245
  language   202–203
  multitude   54, 74, 102, 155, 204
  post-Marxism   102, 174, 229–30, 234, 237
production relations   32, 47–48, 66, 74, 97,
  107, 125, 127–28, 130ff., 136, 159, 182ff.,
  186–87, 191, 193–94, 197, 208, 240
  'changing masters'   49
  debt peonage   48, 127, 136
  deproletarianisation   20–21, 118, 130, 182
    passim, 192, 194ff
  free   38, 130–31, 133, 153, 184, 193ff., 197
  neo-bondage   17, 20–21, 182 *passim*, 191ff.,
    195ff., 243
  patronage   187ff
  serfs   74
  sharecropping   42ff., 46–47, 74, 127, 132
  slavery   18, 35ff., 38–39, 40–41, 43–44, 47,
    50–51, 75, 79, 103, 125ff., 129–30, 132, 141,
    186, 188, 194, 245
  tenants   42 *passim*, 51, 134, 136, 233
  tied housing   47, 136
  unfree   2, 4, 49, 74, 80, 97, 107, 109, 112,
    126–27, 129 *passim*, 141, 159, 182 *passim*,
    191 *passim*, 243

racism   2, 17–18, 25 *passim*, 37, 39ff., 45, 47,
  49–50, 61, 68, 70, 73, 83, 91, 139–40, 228,
  239, 241, 245–46
  anti-foreigner   62
  anti-semitism   28–29, 62

SUBJECT INDEX 277

racism (cont.)
    apartheid   3, 143, 232
    Simon Legree   37–38
    Uncle Tom   37–38
religion   9, 29, 37, 134, 172, 226, 230, 236
    Catholicism   155
    Christianity   28–29, 31, 84
reserve army of labour   XI, 18, 25, 52–53, 55 passim, 67, 69, 76ff., 97–98, 124, 130, 138–39, 146, 150, 157, 167, 175, 179, 182ff., 190, 219, 240–41
    unemployment and   5–6, 18, 59ff., 64–65, 82, 138, 167, 170, 175, 240
revolution   4, 5, 21, 28, 30, 54, 65, 86, 88, 104, 110, 123, 130–31, 152, 155, 168, 179, 221, 229
    French 1789   155
    Hungarian 1848   30
    Green   65, 130, 184
    permanent   4, 110
    Russian 1917   54, 221, 229
Russia   45, 54, 102–103, 179, 212, 221
    USSR   65, 148

social sciences
    anthropology   20, 161, 163 passim, 171, 173, 175, 177, 179–80, 222
    archaeology   162, 201
    fieldwork   20, 124–25, 134, 136, 161ff., 167ff., 176, 180, 188, 191ff., 195, 197, 205, 243–44
    geography   168
    Open Anthropology Cooperative   168
    sociology   10, 19, 49, 84ff., 89, 91–92, 94–95, 97, 99, 116, 221, 242
State, the   1–2, 5ff., 12ff., 18, 29, 39, 44, 59ff., 68–69, 8off., 87, 89, 104, 111, 124, 131, 139, 143, 147, 149, 150, 152–53, 156, 160, 163, 175ff., 187, 190, 202, 216, 221, 228–29, 232–33, 240, 244
    taxation   1, 46, 151, 156, 175

United Kingdom 1945 election   5–6, 63
    2019 election   177
    Brexit   31–32, 76, 79–80, 157, 177–78, 182, 239, 241
    Conservative Party   6, 63, 78–79, 80, 87, 219
    Labour Party   5, 62–63, 80, 87, 143
    Lancashire   173–74
    Liberal Party   62
    Thatcher, Margaret   7, 17, 152, 233
    UKIP   78–79
United States   9–10, 26, 40–41, 52, 56, 75, 86, 92, 128, 152–53, 155, 158, 164, 173, 177
    Civil War   34, 37, 39, 42, 44, 177
    Freedman Bureau   36, 39–40
    Ku Klux Klan   34–35
    lynching   18, 39, 42, 44ff., 48, 50–51, 245
    Northern   35ff., 40–41, 44, 50, 65, 75
    Reagan, Ronald   98, 152
    Reconstruction era   18, 26–27, 34, 50, 241
    Southern   18, 26ff., 34 passim, 45 passim, 103, 125, 129, 239, 241, 245–46

wages   5, 12, 18, 43–44, 46, 48, 50, 55ff., 61, 63–64, 70–71, 76ff., 8off., 137ff., 157, 166, 169, 232–33, 240–41
    subsistence guarantee   187ff
war 1939–45   6, 59–60, 169, 227
    Cold   148, 173
Warriner, Doreen   162
workers   5, 8, 12, 25, 43, 45 passim, 56 passim, 66 passim, 77ff., 81–82, 88, 103, 110, 112, 126ff., 133 passim, 144, 146, 153–54, 157, 173–74, 177–78, 183 passim, 193ff., 203, 211, 215, 217, 227, 232ff., 241 passim. See also capitalism, class
    strikes   56, 64, 75, 137, 232, 240
    trade unions   5–6, 38, 57, 70, 78, 232–33

www.ingramcontent.com/pod-product-compliance
Lightning Source LLC
Chambersburg PA
CBHW070614030426
42337CB00020B/3794